Matrix Analysis
for Scientists & Engineers

Alan J. Laub

University of California
Davis, California

10 9 8 7 6 5 4 3 2 1

Library of Congress Cataloging-in-Publication Data

Laub, Alan J., 1948-
Matrix analysis for scientists and engineers / Alan J. Laub.
p. cm.
Includes bibliographical references and index.
ISBN 0-89871-576-8 (pbk.)
1. Matrices. 2. Mathematical analysis. I. Title.

QA188.L38 2005
512.9'434—dc22

2004059962

About the cover: The original artwork featured on the cover was created by freelance artist Aaron Tallon of Philadelphia, PA. Used by permission.

To my wife, Beverley

(who captivated me in the UBC math library
nearly forty years ago)

Contents

Preface

This book is intended to be used as a text for beginning graduate-level (or even senior-level) students in engineering, the sciences, mathematics, computer science, or computational science who wish to be familar with enough matrix analysis that they are prepared to use its tools and ideas comfortably in a variety of applications. By *matrix analysis* I mean linear algebra and matrix theory together with their intrinsic interaction with and application to linear dynamical systems (systems of linear differential or difference equations). The text can be used in a one-quarter or one-semester course to provide a compact overview of much of the important and useful mathematics that, in many cases, students meant to learn thoroughly as undergraduates, but somehow didn't quite manage to do. Certain topics that may have been treated cursorily in undergraduate courses are treated in more depth and more advanced material is introduced. I have tried throughout to emphasize only the more important and "useful" tools, methods, and mathematical structures. Instructors are encouraged to supplement the book with specific application examples from their own particular subject area.

The choice of topics covered in linear algebra and matrix theory is motivated both by applications and by computational utility and relevance. The concept of matrix factorization is emphasized throughout to provide a foundation for a later course in numerical linear algebra. Matrices are stressed more than abstract vector spaces, although Chapters 2 and 3 do cover some geometric (i.e., basis-free or subspace) aspects of many of the fundamental notions. The books by Meyer [18], Noble and Daniel [20], Ortega [21], and Strang [24] are excellent companion texts for this book. Upon completion of a course based on this text, the student is then well-equipped to pursue, either via formal courses or through self-study, follow-on topics on the computational side (at the level of [7], [11], [23], or [25], for example) or on the theoretical side (at the level of [12], [13], or [16], for example).

Prerequisites for using this text are quite modest: essentially just an understanding of calculus and definitely some previous exposure to matrices and linear algebra. Basic concepts such as determinants, singularity of matrices, eigenvalues and eigenvectors, and positive definite matrices should have been covered at least once, even though their recollection may occasionally be "hazy." However, requiring such material as prerequisite permits the early (but "out-of-order" by conventional standards) introduction of topics such as pseudoinverses and the singular value decomposition (SVD). These powerful and versatile tools can then be exploited to provide a unifying foundation upon which to base subsequent topics. Because tools such as the SVD are not generally amenable to "hand computation," this approach necessarily presupposes the availability of appropriate mathematical software on a digital computer. For this, I highly recommend MATLAB® although other software such as

Mathematica® or Mathcad® is also excellent. Since this text is not intended for a course in numerical linear algebra *per se*, the details of most of the numerical aspects of linear algebra are deferred to such a course.

The presentation of the material in this book is strongly influenced by computational issues for two principal reasons. First, "real-life" problems seldom yield to simple closed-form formulas or solutions. They must generally be solved computationally and it is important to know which types of algorithms can be relied upon and which cannot. Some of the key algorithms of numerical linear algebra, in particular, form the foundation upon which rests virtually all of modern scientific and engineering computation. A second motivation for a computational emphasis is that it provides many of the essential tools for what I call "qualitative mathematics." For example, in an elementary linear algebra course, a set of vectors is either linearly independent or it is not. This is an absolutely fundamental concept. But in most engineering or scientific contexts we want to know more than that. If a set of vectors is linearly independent, how "nearly dependent" are the vectors? If they are linearly dependent, are there "best" linearly independent subsets? These turn out to be much more difficult problems and frequently involve research-level questions when set in the context of the finite-precision, finite-range floating-point arithmetic environment of most modern computing platforms.

Some of the applications of matrix analysis mentioned briefly in this book derive from the modern state-space approach to dynamical systems. State-space methods are now standard in much of modern engineering where, for example, control systems with large numbers of interacting inputs, outputs, and states often give rise to models of very high order that must be analyzed, simulated, and evaluated. The "language" in which such models are conveniently described involves vectors and matrices. It is thus crucial to acquire a working knowledge of the vocabulary and grammar of this language. The tools of matrix analysis are also applied on a daily basis to problems in biology, chemistry, econometrics, physics, statistics, and a wide variety of other fields, and thus the text can serve a rather diverse audience. Mastery of the material in this text should enable the student to read and understand the modern language of matrices used throughout mathematics, science, and engineering.

While prerequisites for this text are modest, and while most material is developed from basic ideas in the book, the student does require a certain amount of what is conventionally referred to as "mathematical maturity." Proofs are given for many theorems. When they are not given explicitly, they are either obvious or easily found in the literature. This is ideal material from which to learn a bit about mathematical proofs and the mathematical maturity and insight gained thereby. It is my firm conviction that such maturity is neither encouraged nor nurtured by relegating the mathematical aspects of applications (for example, linear algebra for elementary state-space theory) to an appendix or introducing it "on-the-fly" when necessary. Rather, one must lay a firm foundation upon which subsequent applications and perspectives can be built in a logical, consistent, and coherent fashion.

I have taught this material for many years, many times at UCSB and twice at UC Davis, and the course has proven to be remarkably successful at enabling students from disparate backgrounds to acquire a quite acceptable level of mathematical maturity and rigor for subsequent graduate studies in a variety of disciplines. Indeed, many students who completed the course, especially the first few times it was offered, remarked afterward that if only they had had this course before they took linear systems, or signal processing, or estimation theory, etc., they would have been able to concentrate on the new ideas they wanted to learn, rather than having to spend time making up for deficiencies in their background in matrices and linear algebra. My fellow instructors, too, realized that by requiring this course as a prerequisite, they no longer had to provide as much time for "review" and could focus instead on the subject at hand. The concept seems to work.

— AJL, June 2004

Chapter 1

Introduction and Review

1.1 Some Notation and Terminology

We begin with a brief introduction to some standard notation and terminology to be used throughout the text. This is followed by a review of some basic notions in matrix analysis and linear algebra.

The following sets appear frequently throughout subsequent chapters:

1. $\mathbb{R}^n$ = the set of n-tuples of real numbers represented as column vectors. Thus, $x \in \mathbb{R}^n$ means

$$x = \begin{bmatrix} x_1 \\ x_2 \\ \vdots \\ x_n \end{bmatrix},$$

   where $x_i \in \mathbb{R}$ for $i \in \underline{n}$.
   Henceforth, the notation $\underline{n}$ denotes the set $\{1, \ldots, n\}$.

   Note: Vectors are **always column vectors**. A row vector is denoted by y^T, where $y \in \mathbb{R}^n$ and the superscript T is the transpose operation. That a vector is always a column vector rather than a row vector is entirely arbitrary, but this convention makes it easy to recognize immediately throughout the text that, e.g., $x^T y$ is a scalar while xy^T is an $n \times n$ matrix.

2. $\mathbb{C}^n$ = the set of n-tuples of complex numbers represented as column vectors.

3. $\mathbb{R}^{m \times n}$ = the set of real (or real-valued) $m \times n$ matrices.

4. $\mathbb{R}_r^{m \times n}$ = the set of real $m \times n$ matrices of rank r. Thus, $\mathbb{R}_n^{n \times n}$ denotes the set of real nonsingular $n \times n$ matrices.

5. $\mathbb{C}^{m \times n}$ = the set of complex (or complex-valued) $m \times n$ matrices.

6. $\mathbb{C}_r^{m \times n}$ = the set of complex $m \times n$ matrices of rank r.

We now classify some of the more familiar "shaped" matrices. A matrix $A \in \mathbb{R}^{n \times n}$ (or $A \in \mathbb{C}^{n \times n}$) is

- **diagonal** if $a_{ij} = 0$ for $i \neq j$.
- **upper triangular** if $a_{ij} = 0$ for $i > j$.
- **lower triangular** if $a_{ij} = 0$ for $i < j$.
- **tridiagonal** if $a_{ij} = 0$ for $|i - j| > 1$.
- **pentadiagonal** if $a_{ij} = 0$ for $|i - j| > 2$.
- **upper Hessenberg** if $a_{ij} = 0$ for $i - j > 1$.
- **lower Hessenberg** if $a_{ij} = 0$ for $j - i > 1$.

Each of the above also has a "block" analogue obtained by replacing scalar components in the respective definitions by block submatrices. For example, if $A \in \mathbb{R}^{n \times n}$, $B \in \mathbb{R}^{n \times m}$, and $C \in \mathbb{R}^{m \times m}$, then the $(m+n) \times (m+n)$ matrix $\left[\begin{smallmatrix} A & B \\ 0 & C \end{smallmatrix}\right]$ is block upper triangular.

The **transpose** of a matrix A is denoted by A^T and is the matrix whose (i, j)th entry is the (j, i)th entry of A, that is, $(A^T)_{ij} = a_{ji}$. Note that if $A \in \mathbb{R}^{m \times n}$, then $A^T \in \mathbb{R}^{n \times m}$. If $A \in \mathbb{C}^{m \times n}$, then its **Hermitian transpose** (or conjugate transpose) is denoted by A^H (or sometimes A^*) and its (i, j)th entry is $(A^H)_{ij} = (\overline{a_{ji}})$, where the bar indicates complex conjugation; i.e., if $z = \alpha + j\beta$ ($j = i = \sqrt{-1}$), then $\bar{z} = \alpha - j\beta$. A matrix A is **symmetric** if $A = A^T$ and **Hermitian** if $A = A^H$. We henceforth adopt the convention that, unless otherwise noted, an equation like $A = A^T$ implies that A is real-valued while a statement like $A = A^H$ implies that A is complex-valued.

Remark 1.1. While $\sqrt{-1}$ is most commonly denoted by i in mathematics texts, j is the more common notation in electrical engineering and system theory. There is some advantage to being conversant with both notations. The notation j is used throughout the text but reminders are placed at strategic locations.

Example 1.2.

1. $A = \begin{bmatrix} 5 & 7 \\ 7 & 2 \end{bmatrix}$ is symmetric (and Hermitian).

2. $A = \begin{bmatrix} 5 & 7+j \\ 7+j & 2 \end{bmatrix}$ is complex-valued symmetric but not Hermitian.

3. $A = \begin{bmatrix} 5 & 7+j \\ 7-j & 2 \end{bmatrix}$ is Hermitian (but not symmetric).

Transposes of block matrices can be defined in an obvious way. For example, it is easy to see that if A_{ij} are appropriately dimensioned subblocks, then

$$\begin{bmatrix} A_{11} & A_{12} \\ A_{21} & A_{22} \end{bmatrix}^T = \begin{bmatrix} A_{11}^T & A_{21}^T \\ A_{12}^T & A_{22}^T \end{bmatrix}.$$

1.2 Matrix Arithmetic

It is assumed that the reader is familiar with the fundamental notions of matrix addition, multiplication of a matrix by a scalar, and multiplication of matrices.

A special case of matrix multiplication occurs when the second matrix is a column vector x, i.e., the matrix-vector product Ax. A very important way to view this product is to interpret it as a weighted sum (linear combination) of the columns of A. That is, suppose

$$A = [a_1, \ldots, a_n] \in \mathbb{R}^{m \times n} \text{ with } a_i \in \mathbb{R}^m \text{ and } x = \begin{bmatrix} x_1 \\ x_2 \\ \vdots \\ x_n \end{bmatrix} \in \mathbb{R}^n.$$

Then

$$Ax = x_1 a_1 + \cdots + x_n a_n \in \mathbb{R}^m.$$

The importance of this interpretation cannot be overemphasized. As a numerical example, take $A = \left[\begin{smallmatrix} 9 & 8 & 7 \\ 6 & 5 & 4 \end{smallmatrix}\right]$, $x = \left[\begin{smallmatrix} 3 \\ 2 \\ 1 \end{smallmatrix}\right]$. Then we can quickly calculate dot products of the rows of A with the column x to find $Ax = \left[\begin{smallmatrix} 50 \\ 32 \end{smallmatrix}\right]$, but this matrix-vector product can also be computed via

$$3 \cdot \begin{bmatrix} 9 \\ 6 \end{bmatrix} + 2 \cdot \begin{bmatrix} 8 \\ 5 \end{bmatrix} + 1 \cdot \begin{bmatrix} 7 \\ 4 \end{bmatrix}.$$

For large arrays of numbers, there can be important computer-architecture-related advantages to preferring the latter calculation method.

For matrix multiplication, suppose $A \in \mathbb{R}^{m \times n}$ and $B = \left[b_1, \ldots, b_p\right] \in \mathbb{R}^{n \times p}$ with $b_i \in \mathbb{R}^n$. Then the matrix product AB can be thought of as above, applied p times:

$$AB = A\left[b_1, \ldots, b_p\right] = \left[Ab_1, \ldots, Ab_p\right] \in \mathbb{R}^{m \times p}.$$

There is also an alternative, but equivalent, formulation of matrix multiplication that appears frequently in the text and is presented below as a theorem. Again, its importance cannot be overemphasized. It is deceptively simple and its full understanding is well rewarded.

Theorem 1.3. *Let $U = [u_1, \ldots, u_n] \in \mathbb{R}^{m \times n}$ with $u_i \in \mathbb{R}^m$ and $V = [v_1, \ldots, v_n] \in \mathbb{R}^{p \times n}$ with $v_i \in \mathbb{R}^p$. Then*

$$UV^T = \sum_{i=1}^{n} u_i v_i^T \in \mathbb{R}^{m \times p}.$$

If matrices C and D are compatible for multiplication, recall that $(CD)^T = D^T C^T$ (or $(CD)^H = D^H C^H$). This gives a dual to the matrix-vector result above. Namely, if $C \in \mathbb{R}^{m \times n}$ has row vectors $c_j^T \in \mathbb{R}^{1 \times n}$, and is premultiplied by a row vector $y^T \in \mathbb{R}^{1 \times m}$, then the product can be written as a weighted linear sum of the rows of C as follows:

$$y^T C = y_1 c_1^T + \cdots + y_m c_m^T \in \mathbb{R}^{1 \times n}.$$

Theorem 1.3 can then also be generalized to its "row dual." The details are left to the reader.

1.3 Inner Products and Orthogonality

For vectors $x, y \in \mathbb{R}^n$, the **Euclidean inner product** (or inner product, for short) of x and y is given by

$$\langle x, y \rangle := x^T y = \sum_{i=1}^{n} x_i y_i .$$

Note that the inner product is a scalar.

If $x, y \in \mathbb{C}^n$, we define their **complex Euclidean inner product** (or inner product, for short) by

$$\langle x, y \rangle_c := x^H y = \sum_{i=1}^{n} \overline{x_i} y_i .$$

Note that $\langle x, y \rangle_c = \overline{\langle y, x \rangle_c}$, i.e., the order in which x and y appear in the complex inner product is important. The more conventional definition of the complex inner product is $\langle x, y \rangle_c = y^H x = \sum_{i=1}^{n} x_i \overline{y_i}$ but throughout the text we prefer the symmetry with the real case.

Example 1.4. Let $x = \begin{bmatrix} 1 \\ j \end{bmatrix}$ and $y = \begin{bmatrix} 1 \\ 2 \end{bmatrix}$. Then

$$\langle x, y \rangle_c = \begin{bmatrix} 1 \\ j \end{bmatrix}^H \begin{bmatrix} 1 \\ 2 \end{bmatrix} = [1 \quad -j] \begin{bmatrix} 1 \\ 2 \end{bmatrix} = 1 - 2j$$

while

$$\langle y, x \rangle_c = \begin{bmatrix} 1 \\ 2 \end{bmatrix}^H \begin{bmatrix} 1 \\ j \end{bmatrix} = [1 \quad 2] \begin{bmatrix} 1 \\ j \end{bmatrix} = 1 + 2j$$

and we see that, indeed, $\langle x, y \rangle_c = \overline{\langle y, x \rangle_c}$.

Note that $x^T x = 0$ if and only if $x = 0$ when $x \in \mathbb{R}^n$ but that this is not true if $x \in \mathbb{C}^n$. What is true in the complex case is that $x^H x = 0$ if and only if $x = 0$. To illustrate, consider the nonzero vector x above. Then $x^T x = 0$ but $x^H x = 2$.

Two nonzero vectors $x, y \in \mathbb{R}^n$ are said to be **orthogonal** if their inner product is zero, i.e., $x^T y = 0$. Nonzero complex vectors are orthogonal if $x^H y = 0$. If x and y are orthogonal and $x^T x = 1$ and $y^T y = 1$, then we say that x and y are **orthonormal**. A matrix $A \in \mathbb{R}^{n \times n}$ is an **orthogonal matrix** if $A^T A = A A^T = I$, where I is the $n \times n$ **identity matrix**. The notation I_n is sometimes used to denote the identity matrix in $\mathbb{R}^{n \times n}$ (or $\mathbb{C}^{n \times n}$). Similarly, a matrix $A \in \mathbb{C}^{n \times n}$ is said to be **unitary** if $A^H A = A A^H = I$. Clearly an orthogonal or unitary matrix has orthonormal rows and orthonormal columns. There is no special name attached to a nonsquare matrix $A \in \mathbb{R}^{m \times n}$ (or $\in \mathbb{C}^{m \times n}$) with orthonormal rows or columns.

1.4 Determinants

It is assumed that the reader is familiar with the basic theory of determinants. For $A \in \mathbb{R}^{n \times n}$ (or $A \in \mathbb{C}^{n \times n}$) we use the notation $\det A$ for the determinant of A. We list below some of the more useful properties of determinants. Note that this is not a minimal set, i.e., several properties are consequences of one or more of the others.

1. If A has a zero row or if any two rows of A are equal, then $\det A = 0$.
2. If A has a zero column or if any two columns of A are equal, then $\det A = 0$.
3. Interchanging two rows of A changes only the sign of the determinant.
4. Interchanging two columns of A changes only the sign of the determinant.
5. Multiplying a row of A by a scalar α results in a new matrix whose determinant is $\alpha \det A$.
6. Multiplying a column of A by a scalar α results in a new matrix whose determinant is $\alpha \det A$.
7. Multiplying a row of A by a scalar and then adding it to another row does not change the determinant.
8. Multiplying a column of A by a scalar and then adding it to another column does not change the determinant.
9. $\det A^T = \det A$ ($\det A^H = \overline{\det A}$ if $A \in \mathbb{C}^{n\times n}$).
10. If A is diagonal, then $\det A = a_{11}a_{22}\cdots a_{nn}$, i.e., $\det A$ is the product of its diagonal elements.
11. If A is upper triangular, then $\det A = a_{11}a_{22}\cdots a_{nn}$.
12. If A is lower triangular, then $\det A = a_{11}a_{22}\cdots a_{nn}$.
13. If A is block diagonal (or block upper triangular or block lower triangular), with square diagonal blocks $A_{11}, A_{22}, \ldots, A_{nn}$ (of possibly different sizes), then $\det A = \det A_{11} \det A_{22} \cdots \det A_{nn}$.
14. If $A, B \in \mathbb{R}^{n\times n}$, then $\det(AB) = \det A \det B$.
15. If $A \in \mathbb{R}_n^{n\times n}$, then $\det(A^{-1}) = \frac{1}{\det A}$.
16. If $A \in \mathbb{R}_n^{n\times n}$ and $D \in \mathbb{R}^{m\times m}$, then $\det \left[\begin{smallmatrix} A & B \\ C & D \end{smallmatrix}\right] = \det A \det(D - CA^{-1}B)$.
    Proof: This follows easily from the block LU factorization

$$\begin{bmatrix} A & B \\ C & D \end{bmatrix} = \begin{bmatrix} I & 0 \\ CA^{-1} & I \end{bmatrix} \begin{bmatrix} A & B \\ 0 & D - CA^{-1}B \end{bmatrix}.$$

17. If $A \in \mathbb{R}^{n\times n}$ and $D \in \mathbb{R}_m^{m\times m}$, then $\det \left[\begin{smallmatrix} A & B \\ C & D \end{smallmatrix}\right] = \det D \det(A - BD^{-1}C)$.
    Proof: This follows easily from the block UL factorization

$$\begin{bmatrix} A & B \\ C & D \end{bmatrix} = \begin{bmatrix} I & BD^{-1} \\ 0 & I \end{bmatrix} \begin{bmatrix} A - BD^{-1}C & 0 \\ C & D \end{bmatrix}.$$

Remark 1.5. The factorization of a matrix A into the product of a unit lower triangular matrix L (i.e., lower triangular with all 1's on the diagonal) and an upper triangular matrix U is called an LU factorization; see, for example, [24]. Another such factorization is UL where U is unit upper triangular and L is lower triangular. The factorizations used above are block analogues of these.

Remark 1.6. The matrix $D - CA^{-1}B$ is called the **Schur complement** of A in $\left[\begin{smallmatrix} A & B \\ C & D \end{smallmatrix}\right]$. Similarly, $A - BD^{-1}C$ is the Schur complement of D in $\left[\begin{smallmatrix} A & B \\ C & D \end{smallmatrix}\right]$.

EXERCISES

1. If $A \in \mathbb{R}^{n \times n}$ and α is a scalar, what is $\det(\alpha A)$? What is $\det(-A)$?

2. If A is orthogonal, what is $\det A$? If A is unitary, what is $\det A$?

3. Let $x, y \in \mathbb{R}^n$. Show that $\det(I - xy^T) = 1 - y^T x$.

4. Let $U_1, U_2, \ldots, U_k \in \mathbb{R}^{n \times n}$ be orthogonal matrices. Show that the product $U = U_1 U_2 \cdots U_k$ is an orthogonal matrix.

5. Let $A \in \mathbb{R}^{n \times n}$. The *trace* of A, denoted $\mathrm{Tr}A$, is defined as the sum of its diagonal elements, i.e., $\mathrm{Tr}A = \sum_{i=1}^{n} a_{ii}$.

   (a) Show that the trace is a linear function; i.e., if $A, B \in \mathbb{R}^{n \times n}$ and $\alpha, \beta \in \mathbb{R}$, then $\mathrm{Tr}(\alpha A + \beta B) = \alpha \mathrm{Tr}A + \beta \mathrm{Tr}B$.

   (b) Show that $\mathrm{Tr}(AB) = \mathrm{Tr}(BA)$, even though in general $AB \neq BA$.

   (c) Let $S \in \mathbb{R}^{n \times n}$ be skew-symmetric, i.e., $S^T = -S$. Show that $\mathrm{Tr}S = 0$. Then either prove the converse or provide a counterexample.

6. A matrix $A \in \mathbb{R}^{n \times n}$ is said to be *idempotent* if $A^2 = A$.

   (a) Show that the matrix $A = \dfrac{1}{2}\begin{bmatrix} 2\cos^2\theta & \sin 2\theta \\ \sin 2\theta & 2\sin^2\theta \end{bmatrix}$ is idempotent for all θ.

   (b) Suppose $A \in \mathbb{R}^{n \times n}$ is idempotent and $A \neq I$. Show that A must be singular.

Chapter 2

Vector Spaces

In this chapter we give a brief review of some of the basic concepts of vector spaces. The emphasis is on finite-dimensional vector spaces, including spaces formed by special classes of matrices, but some infinite-dimensional examples are also cited. An excellent reference for this and the next chapter is [10], where some of the proofs that are not given here may be found.

2.1 Definitions and Examples

Definition 2.1. *A* **field** *is a set* $\mathbb{F}$ *together with two operations* $+, \cdot : \mathbb{F} \times \mathbb{F} \rightarrow \mathbb{F}$ *such that*

(A1) $\alpha + (\beta + \gamma) = (\alpha + \beta) + \gamma$ *for all* $\alpha, \beta, \gamma \in \mathbb{F}$.

(A2) there exists an element $0 \in \mathbb{F}$ *such that* $\alpha + 0 = \alpha$ *for all* $\alpha \in \mathbb{F}$.

(A3) for all $\alpha \in \mathbb{F}$, *there exists an element* $(-\alpha) \in \mathbb{F}$ *such that* $\alpha + (-\alpha) = 0$.

(A4) $\alpha + \beta = \beta + \alpha$ *for all* $\alpha, \beta \in \mathbb{F}$.

(M1) $\alpha \cdot (\beta \cdot \gamma) = (\alpha \cdot \beta) \cdot \gamma$ *for all* $\alpha, \beta, \gamma \in \mathbb{F}$.

(M2) there exists an element $1 \in \mathbb{F}$ *such that* $\alpha \cdot 1 = \alpha$ *for all* $\alpha \in \mathbb{F}$.

(M3) for all $\alpha \in \mathbb{F}$, $\alpha \neq 0$, *there exists an element* $\alpha^{-1} \in \mathbb{F}$ *such that* $\alpha \cdot \alpha^{-1} = 1$.

(M4) $\alpha \cdot \beta = \beta \cdot \alpha$ *for all* $\alpha, \beta \in \mathbb{F}$.

(D) $\alpha \cdot (\beta + \gamma) = \alpha \cdot \beta + \alpha \cdot \gamma$ *for all* $\alpha, \beta, \gamma \in \mathbb{F}$.

Axioms (A1)–(A3) state that $(\mathbb{F}, +)$ is a group and an abelian group if (A4) also holds. Axioms (M1)–(M4) state that $(\mathbb{F} \setminus \{0\}, \cdot)$ is an abelian group.

Generally speaking, when no confusion can arise, the multiplication operator "$\cdot$" is not written explicitly.

Example 2.2.

1. $\mathbb{R}$ with ordinary addition and multiplication is a field.

2. $\mathbb{C}$ with ordinary complex addition and multiplication is a field.

3. $\text{Ra}[x]$ = the **field of rational functions in the indeterminate** x

$$= \left\{ \frac{\alpha_0 + \alpha_1 x + \cdots + \alpha_p x^p}{\beta_0 + \beta_1 x + \cdots + \beta_q x^q} \;:\; \alpha_i, \beta_i \in \mathbb{R} \; ; \; p, q \in \mathbf{Z}^+ \right\},$$

where $\mathbf{Z}^+ = \{0,1,2,\ldots\}$, is a field.

4. $\mathbb{R}_r^{m \times n} = \{ \, m \times n$ matrices of rank r with real coefficients$\}$ is clearly not a field since, for example, (M1) does not hold unless $m = n$. Moreover, $\mathbb{R}_n^{n \times n}$ is not a field either since (M4) does not hold in general (although the other 8 axioms hold).

Definition 2.3. *A* **vector space over a field** $\mathbb{F}$ *is a set* $\mathcal{V}$ *together with two operations* $+ : \mathcal{V} \times \mathcal{V} \to \mathcal{V}$ *and* $\cdot : \mathbb{F} \times \mathcal{V} \to \mathcal{V}$ *such that*

(V1) $(\mathcal{V}, +)$ *is an abelian group.*

(V2) $(\alpha \cdot \beta) \cdot v = \alpha \cdot (\beta \cdot v)$ *for all* $\alpha, \beta \in \mathbb{F}$ *and for all* $v \in \mathcal{V}$.

(V3) $(\alpha + \beta) \cdot v = \alpha \cdot v + \beta \cdot v$ *for all* $\alpha, \beta \in \mathbb{F}$ *and for all* $v \in \mathcal{V}$.

(V4) $\alpha \cdot (v + w) = \alpha \cdot v + \alpha \cdot w$ *for all* $\alpha \in \mathbb{F}$ *and for all* $v, w \in \mathcal{V}$.

(V5) $1 \cdot v = v$ *for all* $v \in \mathcal{V}$ $(1 \in \mathbb{F})$.

A vector space is denoted by $(\mathcal{V}, \mathbb{F})$ *or, when there is no possibility of confusion as to the underlying field, simply by* $\mathcal{V}$.

Remark 2.4. Note that $+$ and $\cdot$ in Definition 2.3 are different from the $+$ and $\cdot$ in Definition 2.1 in the sense of operating on different objects in different sets. In practice, this causes no confusion and the $\cdot$ operator is usually not even written explicitly.

Example 2.5.

1. $(\mathbb{R}^n, \mathbb{R})$ with addition defined by

$$v + w = \begin{bmatrix} v_1 + w_1 \\ \vdots \\ v_n + w_n \end{bmatrix}$$

and scalar multiplication defined by

$$\alpha v = \begin{bmatrix} \alpha v_1 \\ \vdots \\ \alpha v_n \end{bmatrix}$$

is a vector space. Similar definitions hold for $(\mathbb{C}^n, \mathbb{C})$.

2. $(\mathbb{R}^{m\times n}, \mathbb{R})$ is a vector space with addition defined by

$$A + B = \begin{bmatrix} \alpha_{11}+\beta_{11} & \alpha_{12}+\beta_{12} & \cdots & \alpha_{1n}+\beta_{1n} \\ \alpha_{21}+\beta_{21} & \alpha_{22}+\beta_{22} & \cdots & \alpha_{2n}+\beta_{2n} \\ \vdots & \vdots & \cdots & \vdots \\ \alpha_{m1}+\beta_{m1} & \alpha_{m2}+\beta_{m2} & \cdots & \alpha_{mn}+\beta_{mn} \end{bmatrix}$$

and scalar multiplication defined by

$$\gamma A = \begin{bmatrix} \gamma\alpha_{11} & \gamma\alpha_{12} & \cdots & \gamma\alpha_{1n} \\ \gamma\alpha_{21} & \gamma\alpha_{22} & \cdots & \gamma\alpha_{2n} \\ \vdots & \vdots & \cdots & \vdots \\ \gamma\alpha_{m1} & \gamma\alpha_{m2} & \cdots & \gamma\alpha_{mn} \end{bmatrix}.$$

3. Let $(\mathcal{V}, \mathbb{F})$ be an arbitrary vector space and $\mathcal{D}$ be an arbitrary set. Let $\Phi(\mathcal{D}, \mathcal{V})$ be the set of functions f mapping $\mathcal{D}$ to $\mathcal{V}$. Then $\Phi(\mathcal{D}, \mathcal{V})$ is a vector space with addition defined by

$$(f+g)(d) = f(d) + g(d) \quad \text{for all } d \in \mathcal{D} \text{ and for all } f, g \in \Phi$$

and scalar multiplication defined by

$$(\alpha f)(d) = \alpha f(d) \quad \text{for all } \alpha \in \mathbb{F}, \text{ for all } d \in \mathcal{D}, \text{ and for all } f \in \Phi.$$

Special Cases:

(a) $\mathcal{D} = [t_0, t_1]$, $(\mathcal{V}, \mathbb{F}) = (\mathbb{R}^n, \mathbb{R})$, and the functions are piecewise continuous $=: (\mathbf{PC}[t_0, t_1])^n$ or continuous $=: (\mathbf{C}[t_0, t_1])^n$.

(b) $\mathcal{D} = [t_0, +\infty)$, $(\mathcal{V}, \mathbb{F}) = (\mathbb{R}^n, \mathbb{R})$, etc.

4. Let $A \in \mathbb{R}^{n\times n}$. Then $\{x(t) : \dot{x}(t) = Ax(t)\}$ is a vector space (of dimension n).

2.2 Subspaces

Definition 2.6. *Let $(\mathcal{V}, \mathbb{F})$ be a vector space and let $\mathcal{W} \subseteq \mathcal{V}$, $\mathcal{W} \neq \emptyset$. Then $(\mathcal{W}, \mathbb{F})$ is a* **subspace** *of $(\mathcal{V}, \mathbb{F})$ if and only if $(\mathcal{W}, \mathbb{F})$ is itself a vector space or, equivalently, if and only if $(\alpha w_1 + \beta w_2) \in \mathcal{W}$ for all $\alpha, \beta \in \mathbb{F}$ and for all $w_1, w_2 \in \mathcal{W}$.*

Remark 2.7. The latter characterization of a subspace is often the easiest way to check or prove that something is indeed a subspace (or vector space); i.e., verify that the set in question is closed under addition and scalar multiplication. Note, too, that since $0 \in \mathbb{F}$, this implies that the zero vector must be in any subspace.

Notation: When the underlying field is understood, we write $\mathcal{W} \subseteq \mathcal{V}$, and the symbol $\subseteq$, when used with vector spaces, is henceforth understood to mean "is a subspace of." The less restrictive meaning "is a subset of" is specifically flagged as such.

Example 2.8.

1. Consider $(\mathcal{V}, \mathbb{F}) = (\mathbb{R}^{n\times n}, \mathbb{R})$ and let $\mathcal{W} = \{A \in \mathbb{R}^{n\times n} : A \text{ is symmetric}\}$. Then $\mathcal{W} \subseteq \mathcal{V}$.

   Proof: Suppose A_1, A_2 are symmetric. Then it is easily shown that $\alpha A_1 + \beta A_2$ is symmetric for all $\alpha, \beta \in \mathbb{R}$.

2. Let $\mathcal{W} = \{A \in \mathbb{R}^{n\times n} : A \text{ is orthogonal}\}$. Then $\mathcal{W}$ is *not* a subspace of $\mathbb{R}^{n\times n}$.

3. Consider $(\mathcal{V}, \mathbb{F}) = (\mathbb{R}^2, \mathbb{R})$ and for each $v \in \mathbb{R}^2$ of the form $v = \left[\begin{smallmatrix} v_1 \\ v_2 \end{smallmatrix}\right]$ identify v_1 with the x-coordinate in the plane and v_2 with the y-coordinate. For $\alpha, \beta \in \mathbb{R}$, define

$$\mathcal{W}_{\alpha,\beta} = \left\{ v : v = \begin{bmatrix} c \\ \alpha c + \beta \end{bmatrix} ; c \in \mathbb{R} \right\}.$$

   Then $\mathcal{W}_{\alpha,\beta}$ is a subspace of $\mathcal{V}$ if and only if $\beta = 0$. As an interesting exercise, sketch $\mathcal{W}_{2,1}$, $\mathcal{W}_{2,0}$, $\mathcal{W}_{\frac{1}{2},1}$, and $\mathcal{W}_{\frac{1}{2},0}$. Note, too, that the vertical line through the origin (i.e., $\alpha = \infty$) is also a subspace.

   All lines through the origin are subspaces. Shifted subspaces $\mathcal{W}_{\alpha,\beta}$ with $\beta \neq 0$ are called **linear varieties**.

Henceforth, we drop the explicit dependence of a vector space on an underlying field. Thus, $\mathcal{V}$ usually denotes a vector space with the underlying field generally being $\mathbb{R}$ unless explicitly stated otherwise.

Definition 2.9. *If $\mathcal{R}$ and $\mathcal{S}$ are vector spaces (or subspaces), then $\mathcal{R} = \mathcal{S}$ if and only if $\mathcal{R} \subseteq \mathcal{S}$* **and** *$\mathcal{S} \subseteq \mathcal{R}$.*

Note: To prove two vector spaces are equal, one usually proves the two inclusions separately: An arbitrary $r \in \mathcal{R}$ is shown to be an element of $\mathcal{S}$ and then an arbitrary $s \in \mathcal{S}$ is shown to be an element of $\mathcal{R}$.

2.3 Linear Independence

Let $X = \{v_1, v_2, \ldots\}$ be a nonempty collection of vectors v_i in some vector space $\mathcal{V}$.

Definition 2.10. *X is a* **linearly dependent** *set of vectors if and only if there exist k distinct elements $v_1, \ldots, v_k \in X$ and scalars $\alpha_1, \ldots, \alpha_k$ not all zero such that*

$$\alpha_1 v_1 + \cdots + \alpha_k v_k = 0.$$

X is a **linearly independent** *set of vectors if and only if for* **any** *collection of k distinct elements $v_1, \ldots, v_k$ of X and for any scalars $\alpha_1, \ldots, \alpha_k$,*

$$\alpha_1 v_1 + \cdots + \alpha_k v_k = 0 \textit{ implies } \alpha_1 = 0, \ldots, \alpha_k = 0.$$

Example 2.11.

1. Let $\mathcal{V} = \mathbb{R}^3$. Then $\left\{ \begin{bmatrix} 1 \\ 2 \\ 3 \end{bmatrix}, \begin{bmatrix} 4 \\ 5 \\ 6 \end{bmatrix}, \begin{bmatrix} 7 \\ 8 \\ 8 \end{bmatrix} \right\}$ is a linearly independent set. Why?

   However, $\left\{ \begin{bmatrix} 1 \\ 2 \\ 3 \end{bmatrix}, \begin{bmatrix} 4 \\ 5 \\ 6 \end{bmatrix}, \begin{bmatrix} 2 \\ 1 \\ 0 \end{bmatrix} \right\}$ is a linearly dependent set

   (since $2v_1 - v_2 + v_3 = 0$).

2. Let $A \in \mathbb{R}^{n \times n}$ and $B \in \mathbb{R}^{n \times m}$. Then consider the rows of $e^{tA}B$ as vectors in $\mathbf{C}^m[t_0, t_1]$ (recall that e^{tA} denotes the matrix exponential, which is discussed in more detail in Chapter 11). Independence of these vectors turns out to be equivalent to a concept called *controllability*, to be studied further in what follows.

Let $v_i \in \mathbb{R}^n$, $i \in \underline{k}$, and consider the matrix $V = [v_1, \ldots, v_k] \in \mathbb{R}^{n \times k}$. The linear dependence of this set of vectors is equivalent to the existence of a nonzero vector $a \in \mathbb{R}^k$ such that $Va = 0$. An equivalent condition for linear dependence is that the $k \times k$ matrix $V^T V$ is singular. If the set of vectors is independent, and there exists $a \in \mathbb{R}^k$ such that $Va = 0$, then $a = 0$. An equivalent condition for linear independence is that the matrix $V^T V$ is nonsingular.

Definition 2.12. *Let $X = \{v_1, v_2, \ldots\}$ be a collection of vectors $v_i \in \mathcal{V}$. Then the* **span** *of X is defined as*

$$\begin{aligned} \mathrm{Sp}(X) &= \mathrm{Sp}\{v_1, v_2, \ldots\} \\ &= \{v : v = \alpha_1 v_1 + \cdots + \alpha_k v_k \; ; \; \alpha_i \in \mathbb{F}, v_i \in X, k \in \mathbf{N}\}, \end{aligned}$$

where $\mathbf{N} = \{1, 2, \ldots\}$.

Example 2.13. Let $\mathcal{V} = \mathbb{R}^n$ and define

$$e_1 = \begin{bmatrix} 1 \\ 0 \\ 0 \\ \vdots \\ 0 \end{bmatrix}, e_2 = \begin{bmatrix} 0 \\ 1 \\ 0 \\ \vdots \\ 0 \end{bmatrix}, \cdots, e_n = \begin{bmatrix} 0 \\ 0 \\ 0 \\ \vdots \\ 1 \end{bmatrix}.$$

Then $\mathrm{Sp}\{e_1, e_2, \ldots, e_n\} = \mathbb{R}^n$.

Definition 2.14. *A set of vectors X is a* **basis** *for $\mathcal{V}$ if and only if*

1. *X is a linearly independent set (of* **basis vectors***), and*

2. *$\mathrm{Sp}(X) = \mathcal{V}$.*

Example 2.15. $\{e_1, \ldots, e_n\}$ is a basis for $\mathbb{R}^n$ (sometimes called the **natural basis**).

Now let $b_1, \ldots, b_n$ be a basis (with a specific order associated with the basis vectors) for $\mathcal{V}$. Then for all $v \in \mathcal{V}$ there exists a unique n-tuple $\{\xi_1, \ldots, \xi_n\}$ such that

$$v = \xi_1 b_1 + \cdots + \xi_n b_n = Bx,$$

where

$$B = [b_1, \ldots, b_n], \quad x = \begin{bmatrix} \xi_1 \\ \vdots \\ \xi_n \end{bmatrix}.$$

Definition 2.16. *The scalars $\{\xi_i\}$ are called the* **components** (*or sometimes the* **coordinates**) *of v with respect to the basis $\{b_1, \ldots, b_n\}$ and are unique. We say that the vector x of components* **represents** *the vector v with respect to the basis B.*

Example 2.17. In $\mathbb{R}^n$,

$$v = \begin{bmatrix} v_1 \\ \vdots \\ v_n \end{bmatrix} = v_1 e_1 + v_2 e_2 + \cdots + v_n e_n.$$

We can also determine components of v with respect to another basis. For example, while

$$\begin{bmatrix} 1 \\ 2 \end{bmatrix} = 1 \cdot e_1 + 2 \cdot e_2,$$

with respect to the basis

$$\left\{ \begin{bmatrix} -1 \\ 2 \end{bmatrix}, \begin{bmatrix} 1 \\ -1 \end{bmatrix} \right\}$$

we have

$$\begin{bmatrix} 1 \\ 2 \end{bmatrix} = 3 \cdot \begin{bmatrix} -1 \\ 2 \end{bmatrix} + 4 \cdot \begin{bmatrix} 1 \\ -1 \end{bmatrix}.$$

To see this, write

$$\begin{aligned} \begin{bmatrix} 1 \\ 2 \end{bmatrix} &= x_1 \cdot \begin{bmatrix} -1 \\ 2 \end{bmatrix} + x_2 \cdot \begin{bmatrix} 1 \\ -1 \end{bmatrix} \\ &= \begin{bmatrix} -1 & 1 \\ 2 & -1 \end{bmatrix} \begin{bmatrix} x_1 \\ x_2 \end{bmatrix}. \end{aligned}$$

Then

$$\begin{bmatrix} x_1 \\ x_2 \end{bmatrix} = \begin{bmatrix} -1 & 1 \\ 2 & -1 \end{bmatrix}^{-1} \begin{bmatrix} 1 \\ 2 \end{bmatrix} = \begin{bmatrix} 3 \\ 4 \end{bmatrix}.$$

Theorem 2.18. *The number of elements in a basis of a vector space is independent of the particular basis considered.*

Definition 2.19. *If a basis X for a vector space $\mathcal{V}(\neq \emptyset)$ has n elements, $\mathcal{V}$ is said to be n-***dimensional** *or have* **dimension** *n and we write* $\dim(\mathcal{V}) = n$ *or* $\dim \mathcal{V} = n$. *For* *consistency, and because the* 0 *vector is in any vector space, we define* $\dim(0) = 0$. *A vector space* $\mathcal{V}$ *is* **finite-dimensional** *if there exists a basis* X *with* $n < +\infty$ *elements; otherwise,* $\mathcal{V}$ *is* **infinite-dimensional**.

Thus, Theorem 2.18 says that $\dim(\mathcal{V})$ = the number of elements in a basis.

Example 2.20.

1. $\dim(\mathbb{R}^n) = n$.

2. $\dim(\mathbb{R}^{m \times n}) = mn$.

   Note: Check that a basis for $\mathbb{R}^{m \times n}$ is given by the mn matrices E_{ij}; $i \in \underline{m}$, $j \in \underline{n}$, where E_{ij} is a matrix all of whose elements are 0 except for a 1 in the (i, j)th location. The collection of E_{ij} matrices can be called the "natural basis matrices."

3. $\dim(\mathbf{C}[t_0, t_1]) = +\infty$.

4. $\dim\{A \in \mathbb{R}^{n \times n} : A = A^T\} = \frac{1}{2}n(n+1)$.
   (To see why, determine $\frac{1}{2}n(n+1)$ *symmetric* basis matrices.)

5. $\dim\{A \in \mathbb{R}^{n \times n} : A$ is upper (lower) triangular$\} = \frac{1}{2}n(n+1)$.

2.4 Sums and Intersections of Subspaces

Definition 2.21. *Let* $(\mathcal{V}, \mathbb{F})$ *be a vector space and let* $\mathcal{R}, \mathcal{S} \subseteq \mathcal{V}$. *The* **sum** *and* **intersection** *of* $\mathcal{R}$ *and* $\mathcal{S}$ *are defined respectively by:*

1. $\mathcal{R} + \mathcal{S} = \{r + s : r \in \mathcal{R}, s \in \mathcal{S}\}$.

2. $\mathcal{R} \cap \mathcal{S} = \{v : v \in \mathcal{R}$ *and* $v \in \mathcal{S}\}$.

Theorem 2.22.

1. $\mathcal{R} + \mathcal{S} \subseteq \mathcal{V}$ *(in general,* $\mathcal{R}_1 + \cdots + \mathcal{R}_k =: \sum_{i=1}^{k} \mathcal{R}_i \subseteq \mathcal{V}$, *for finite* k*).*

2. $\mathcal{R} \cap \mathcal{S} \subseteq \mathcal{V}$ *(in general,* $\bigcap_{\alpha \in A} \mathcal{R}_\alpha \subseteq \mathcal{V}$ *for an arbitrary index set* A*).*

Remark 2.23. The union of two subspaces, $\mathcal{R} \cup \mathcal{S}$, is not necessarily a subspace.

Definition 2.24. $\mathcal{T} = \mathcal{R} \oplus \mathcal{S}$ *is the* **direct sum** *of* $\mathcal{R}$ *and* $\mathcal{S}$ *if*

1. $\mathcal{R} \cap \mathcal{S} = 0$, *and*

2. $\mathcal{R} + \mathcal{S} = \mathcal{T}$ *(in general,* $\mathcal{R}_i \cap (\sum_{j \neq i} \mathcal{R}_j) = 0$ *and* $\sum_i \mathcal{R}_i = \mathcal{T}$*).*

The subspaces $\mathcal{R}$ *and* $\mathcal{S}$ *are said to be* **complements** *of each other in* $\mathcal{T}$.

Remark 2.25. The complement of $\mathcal{R}$ (or $\mathcal{S}$) is not unique. For example, consider $\mathcal{V} = \mathbb{R}^2$ and let $\mathcal{R}$ be any line through the origin. Then any other distinct line through the origin is a complement of $\mathcal{R}$. Among all the complements there is a unique one **orthogonal** to $\mathcal{R}$. We discuss more about orthogonal complements elsewhere in the text.

Theorem 2.26. *Suppose $\mathcal{T} = \mathcal{R} \oplus \mathcal{S}$. Then*

1. *every $t \in \mathcal{T}$ can be written uniquely in the form $t = r + s$ with $r \in \mathcal{R}$ and $s \in \mathcal{S}$.*

2. $\dim(\mathcal{T}) = \dim(\mathcal{R}) + \dim(\mathcal{S})$.

Proof: To prove the first part, suppose an arbitrary vector $t \in \mathcal{T}$ can be written in two ways as $t = r_1 + s_1 = r_2 + s_2$, where $r_1, r_2 \in \mathcal{R}$ and $s_1, s_2 \in \mathcal{S}$. Then $r_1 - r_2 = s_2 - s_1$. But $r_1 - r_2 \in \mathcal{R}$ and $s_2 - s_1 \in \mathcal{S}$. Since $\mathcal{R} \cap \mathcal{S} = 0$, we must have $r_1 = r_2$ and $s_1 = s_2$ from which uniqueness follows.

The statement of the second part is a special case of the next theorem. □

Theorem 2.27. *For arbitrary subspaces $\mathcal{R}$, $\mathcal{S}$ of a vector space $\mathcal{V}$,*

$$\dim(\mathcal{R} + \mathcal{S}) = \dim(\mathcal{R}) + \dim(\mathcal{S}) - \dim(\mathcal{R} \cap \mathcal{S}).$$

Example 2.28. Let $\mathcal{U}$ be the subspace of upper triangular matrices in $\mathbb{R}^{n \times n}$ and let $\mathcal{L}$ be the subspace of lower triangular matrices in $\mathbb{R}^{n \times n}$. Then it may be checked that $\mathcal{U} + \mathcal{L} = \mathbb{R}^{n \times n}$ while $\mathcal{U} \cap \mathcal{L}$ is the set of diagonal matrices in $\mathbb{R}^{n \times n}$. Using the fact that dim {diagonal matrices} = n, together with Examples 2.20.2 and 2.20.5, one can easily verify the validity of the formula given in Theorem 2.27.

Example 2.29. Let $(\mathcal{V}, \mathbb{F}) = (\mathbb{R}^{n \times n}, \mathbb{R})$, let $\mathcal{R}$ be the set of skew-symmetric matrices in $\mathbb{R}^{n \times n}$, and let $\mathcal{S}$ be the set of symmetric matrices in $\mathbb{R}^{n \times n}$. Then $\mathcal{V} = \mathcal{R} \oplus \mathcal{S}$.

Proof: This follows easily from the fact that any $A \in \mathbb{R}^{n \times n}$ can be written in the form

$$A = \frac{1}{2}(A + A^T) + \frac{1}{2}(A - A^T).$$

The first matrix on the right-hand side above is in $\mathcal{S}$ while the second is in $\mathcal{R}$.

EXERCISES

1. Suppose $\{v_1, \ldots, v_k\}$ is a linearly dependent set. Then show that one of the vectors must be a linear combination of the others.

2. Let $x_1, x_2, \ldots, x_k \in \mathbb{R}^n$ be nonzero mutually orthogonal vectors. Show that $\{x_1, \ldots, x_k\}$ must be a linearly independent set.

3. Let $v_1, \ldots, v_n$ be orthonormal vectors in $\mathbb{R}^n$. Show that $Av_1, \ldots, Av_n$ are also orthonormal if and only if $A \in \mathbb{R}^{n \times n}$ is orthogonal.

4. Consider the vectors $v_1 = [2 \;\; 1]^T$ and $v_2 = [3 \;\; 1]^T$. Prove that v_1 and v_2 form a basis for $\mathbb{R}^2$. Find the components of the vector $v = [4 \;\; 1]^T$ with respect to this basis.

5. Let $\mathcal{P}$ denote the set of polynomials of degree less than or equal to two of the form $p_0 + p_1 x + p_2 x^2$, where $p_0, p_1, p_2 \in \mathbb{R}$. Show that $\mathcal{P}$ is a vector space over $\mathbb{R}$. Show that the polynomials $1, x$, and $2x^2 - 1$ are a basis for $\mathcal{P}$. Find the components of the polynomial $2 + 3x + 4x^2$ with respect to this basis.

6. Prove Theorem 2.22 (for the case of two subspaces $\mathcal{R}$ and $\mathcal{S}$ only).

7. Let $\mathcal{P}_n$ denote the vector space of polynomials of degree less than or equal to n, and of the form $p(x) = p_0 + p_1 x + \cdots + p_n x^n$, where the coefficients p_i are all real. Let $\mathcal{P}_E$ denote the subspace of all even polynomials in $\mathcal{P}_n$, i.e., those that satisfy the property $p(-x) = p(x)$. Similarly, let $\mathcal{P}_O$ denote the subspace of all odd polynomials, i.e., those satisfying $p(-x) = -p(x)$. Show that $\mathcal{P}_n = \mathcal{P}_E \oplus \mathcal{P}_O$.

8. Repeat Example 2.28 using instead the two subspaces $\mathcal{T}$ of tridiagonal matrices and $\mathcal{U}$ of upper triangular matrices.

Chapter 3

Linear Transformations

3.1 Definition and Examples

We begin with the basic definition of a linear transformation (or linear map, linear function, or linear operator) between two vector spaces.

Definition 3.1. *Let $(\mathcal{V}, \mathbb{F})$ and $(\mathcal{W}, \mathbb{F})$ be vector spaces. Then $\mathcal{L} : \mathcal{V} \rightarrow \mathcal{W}$ is a* **linear transformation** *if and only if*

$$\mathcal{L}(\alpha v_1 + \beta v_2) = \alpha \mathcal{L} v_1 + \beta \mathcal{L} v_2 \quad \text{for all } \alpha, \beta \in \mathbb{F} \text{ and for all } v_1, v_2 \in \mathcal{V}.$$

The vector space $\mathcal{V}$ is called the **domain** *of the transformation $\mathcal{L}$ while $\mathcal{W}$, the space into which it maps, is called the* **co-domain**.

Example 3.2.

1. Let $\mathbb{F} = \mathbb{R}$ and take $\mathcal{V} = \mathcal{W} = \mathbf{PC}[t_0, +\infty)$.
   Define $\mathcal{L} : \mathbf{PC}[t_0, +\infty) \rightarrow \mathbf{PC}[t_0, +\infty)$ by

$$v(t) \mapsto w(t) = (\mathcal{L}v)(t) = \int_{t_0}^{t} e^{-(t-\tau)} v(\tau)\, d\tau.$$

2. Let $\mathbb{F} = \mathbb{R}$ and take $\mathcal{V} = \mathcal{W} = \mathbb{R}^{m \times n}$. Fix $M \in \mathbb{R}^{m \times m}$.
   Define $\mathcal{L} : \mathbb{R}^{m \times n} \rightarrow \mathbb{R}^{m \times n}$ by

$$X \mapsto Y = \mathcal{L}X = MX.$$

3. Let $\mathbb{F} = \mathbb{R}$ and take $\mathcal{V} = \mathcal{P}^n = \{p(x) = \alpha_0 + \alpha_1 x + \cdots + \alpha_n x^n \ : \ \alpha_i \in \mathbb{R}\}$ and $\mathcal{W} = \mathcal{P}^{n-1}$.
   Define $\mathcal{L} : \mathcal{V} \rightarrow \mathcal{W}$ by $\mathcal{L}p = p'$, where $'$ denotes differentiation with respect to x.

3.2 Matrix Representation of Linear Transformations

Linear transformations between vector spaces with specific bases can be represented conveniently in matrix form. Specifically, suppose $\mathcal{L} : (\mathcal{V}, \mathbb{F}) \to (\mathcal{W}, \mathbb{F})$ is linear and further suppose that $\{v_i,\ i \in \underline{n}\}$ and $\{w_j,\ j \in \underline{m}\}$ are bases for $\mathcal{V}$ and $\mathcal{W}$, respectively. Then the ith column of $A = \operatorname{Mat} \mathcal{L}$ (the matrix representation of $\mathcal{L}$ with respect to the given bases for $\mathcal{V}$ and $\mathcal{W}$) is the representation of $\mathcal{L}v_i$ with respect to $\{w_j,\ j \in \underline{m}\}$. In other words,

$$A = \begin{bmatrix} \alpha_{11} & \cdots & \alpha_{1n} \\ \vdots & & \vdots \\ \alpha_{m1} & \cdots & \alpha_{mn} \end{bmatrix} \in \mathbb{R}^{m \times n}$$

represents $\mathcal{L}$ since

$$\begin{aligned} \mathcal{L}v_i &= \alpha_{1i} w_1 + \cdots + \alpha_{mi} w_m \\ &= W a_i, \end{aligned}$$

where $W = [w_1, \ldots, w_m]$ and

$$a_i = \begin{bmatrix} \alpha_{1i} \\ \vdots \\ \alpha_{mi} \end{bmatrix}$$

is the ith column of A. Note that $A = \operatorname{Mat} \mathcal{L}$ depends on the particular bases for $\mathcal{V}$ and $\mathcal{W}$. This could be reflected by subscripts, say, in the notation, but this is usually not done.

The action of $\mathcal{L}$ on an arbitrary vector $v \in \mathcal{V}$ is uniquely determined (by linearity) by its action on a basis. Thus, if $v = \xi_1 v_1 + \cdots + \xi_n v_n = Vx$ (where v, and hence x, is arbitrary), then

$$\begin{aligned} \mathcal{L}Vx = \mathcal{L}v &= \xi_1 \mathcal{L}v_1 + \cdots + \xi_n \mathcal{L}v_n \\ &= \xi_1 W a_1 + \cdots + \xi_n W a_n \\ &= WAx. \end{aligned}$$

Thus, $\mathcal{L}V = WA$ since x was arbitrary.

When $\mathcal{V} = \mathbb{R}^n$, $\mathcal{W} = \mathbb{R}^m$ and $\{v_i,\ i \in \underline{n}\}$, $\{w_j,\ j \in \underline{m}\}$ are the usual (natural) bases, the equation $\mathcal{L}V = WA$ becomes simply $\mathcal{L} = A$. We thus commonly identify A as a linear transformation with its matrix representation, i.e.,

$$A \in \mathbb{R}^{m \times n} \Longleftrightarrow A : \mathbb{R}^n \to \mathbb{R}^m \Longleftrightarrow \mathbb{R}^n \xrightarrow{A} \mathbb{R}^m.$$

Thinking of A both as a matrix and as a linear transformation from $\mathbb{R}^n$ to $\mathbb{R}^m$ usually causes no confusion. Change of basis then corresponds naturally to appropriate matrix multiplication.

3.3 Composition of Transformations

Consider three vector spaces $\mathcal{U}$, $\mathcal{V}$, and $\mathcal{W}$ and transformations B from $\mathcal{U}$ to $\mathcal{V}$ and A from $\mathcal{V}$ to $\mathcal{W}$. Then we can define a new transformation C as follows:

$$\mathcal{W} \xleftarrow{A} \mathcal{V} \xleftarrow{B} \mathcal{U} \qquad \underbrace{\longleftarrow}_{C}$$

The above diagram illustrates the composition of transformations $C = AB$. Note that in most texts, the arrows above are reversed as follows:

$$\mathcal{U} \xrightarrow{B} \mathcal{V} \xrightarrow{A} \mathcal{W} \qquad \underbrace{\longrightarrow}_{C}$$

However, it might be useful to prefer the former since the transformations A and B appear in the same order in both the diagram and the equation. If $\dim \mathcal{U} = p$, $\dim \mathcal{V} = n$, and $\dim \mathcal{W} = m$, and if we associate matrices with the transformations in the usual way, then composition of transformations corresponds to standard matrix multiplication. That is, we have $\underset{m \times p}{C} = \underset{m \times n}{A} \; \underset{n \times p}{B}$. The above is sometimes expressed componentwise by the formula

$$c_{ij} = \sum_{k=1}^{n} a_{ik} b_{kj}.$$

Two Special Cases:

Inner Product: Let $x, y \in \mathbb{R}^n$. Then their inner product is the scalar

$$x^T y = \sum_{i=1}^{n} x_i y_i.$$

Outer Product: Let $x \in \mathbb{R}^m$, $y \in \mathbb{R}^n$. Then their outer product is the $m \times n$ matrix

$$xy^T = \begin{bmatrix} x_1 y_1 & \cdots & x_1 y_n \\ \vdots & & \vdots \\ x_m y_1 & \cdots & x_m y_n \end{bmatrix}.$$

Note that any **rank-one matrix** $A \in \mathbb{R}^{m \times n}$ can be written in the form $A = xy^T$ above (or xy^H if $A \in \mathbb{C}^{m \times n}$). A rank-one symmetric matrix can be written in the form xx^T (or xx^H).

3.4 Structure of Linear Transformations

Let $A : \mathcal{V} \to \mathcal{W}$ be a linear transformation.

Definition 3.3. *The* **range** *of A, denoted $\mathcal{R}(A)$, is the set $\{w \in \mathcal{W} : w = Av$ for some $v \in \mathcal{V}\}$. Equivalently, $\mathcal{R}(A) = \{Av \ : \ v \in \mathcal{V}\}$. The range of A is also known as the* **image** *of A and denoted $Im(A)$.*

The **nullspace** *of A, denoted $\mathcal{N}(A)$, is the set $\{v \in \mathcal{V} \ : \ Av = 0\}$. The nullspace of A is also known as the* **kernel** *of A and denoted $Ker\,(A)$.*

Theorem 3.4. *Let $A : \mathcal{V} \to \mathcal{W}$ be a linear transformation. Then*

1. $\mathcal{R}(A) \subseteq \mathcal{W}$.
2. $\mathcal{N}(A) \subseteq \mathcal{V}$.

Note that $\mathcal{N}(A)$ and $\mathcal{R}(A)$ are, in general, subspaces of different spaces.

Theorem 3.5. *Let $A \in \mathbb{R}^{m \times n}$. If A is written in terms of its columns as $A = [a_1, \ldots, a_n]$, then*

$$\mathcal{R}(A) = \mathrm{Sp}\{a_1, \ldots, a_n\}\ .$$

Proof: The proof of this theorem is easy, essentially following immediately from the definition. □

Remark 3.6. Note that in Theorem 3.5 and throughout the text, the same symbol (A) is used to denote both a linear transformation and its matrix representation with respect to the usual (natural) bases. See also the last paragraph of Section 3.2.

Definition 3.7. *Let $\{v_1, \ldots, v_k\}$ be a set of nonzero vectors $v_i \in \mathbb{R}^n$. The set is said to be* **orthogonal** *if $v_i^T v_j = 0$ for $i \neq j$ and* **orthonormal** *if $v_i^T v_j = \delta_{ij}$, where δ_{ij} is the* **Kronecker delta** *defined by*

$$\delta_{ij} = \begin{cases} 1 & \text{if } i = j, \\ 0 & \text{if } i \neq j. \end{cases}$$

Example 3.8.

1. $\left\{ \begin{bmatrix} 3 \\ 5 \\ 7 \end{bmatrix}, \begin{bmatrix} -4 \\ 1 \\ 1 \end{bmatrix} \right\}$ is an orthogonal set.
2. $\left\{ \begin{bmatrix} 2/3 \\ 2/3 \\ 1/3 \end{bmatrix}, \begin{bmatrix} 1/\sqrt{2} \\ -1/\sqrt{2} \\ 0 \end{bmatrix} \right\}$ is an orthonormal set.
3. If $\{v_1, \ldots, v_k\}$ with $v_i \in \mathbb{R}^n$ is an orthogonal set, then $\left\{ \frac{v_1}{\sqrt{v_1^T v_1}}, \ldots, \frac{v_k}{\sqrt{v_k^T v_k}} \right\}$ is an orthonormal set.

Definition 3.9. *Let $\mathcal{S} \subseteq \mathbb{R}^n$. Then the* **orthogonal complement** *of $\mathcal{S}$ is defined as the set*

$$\mathcal{S}^\perp = \{v \in \mathbb{R}^n \ : \ v^T s = 0 \ \textit{for all} \ s \in \mathcal{S}\}.$$

Example 3.10. Let

$$\mathcal{S} = \mathrm{Sp}\left\{ \begin{bmatrix} 3 \\ 5 \\ 7 \end{bmatrix}, \begin{bmatrix} -4 \\ 1 \\ 1 \end{bmatrix} \right\}.$$

Then it can be shown that

$$\mathcal{S}^\perp = \mathrm{Sp}\left\{ \begin{bmatrix} 2 \\ 31 \\ -23 \end{bmatrix} \right\}.$$

Working from the definition, the computation involved is simply to find all nontrivial (i.e., nonzero) solutions of the system of equations

$$\begin{aligned} 3x_1 + 5x_2 + 7x_3 &= 0, \\ -4x_1 + \ x_2 + \ x_3 &= 0. \end{aligned}$$

Note that there is nothing special about the two vectors in the basis defining $\mathcal{S}$ being orthogonal. Any set of vectors will do, including dependent spanning vectors (which would, of course, then give rise to redundant equations).

Theorem 3.11. *Let $\mathcal{R}, \mathcal{S} \subseteq \mathbb{R}^n$. Then*

1. $\mathcal{S}^\perp \subseteq \mathbb{R}^n$.
2. $\mathcal{S} \oplus \mathcal{S}^\perp = \mathbb{R}^n$.
3. $(\mathcal{S}^\perp)^\perp = \mathcal{S}$.
4. $\mathcal{R} \subseteq \mathcal{S}$ *if and only if* $\mathcal{S}^\perp \subseteq \mathcal{R}^\perp$.
5. $(\mathcal{R} + \mathcal{S})^\perp = \mathcal{R}^\perp \cap \mathcal{S}^\perp$.
6. $(\mathcal{R} \cap \mathcal{S})^\perp = \mathcal{R}^\perp + \mathcal{S}^\perp$.

Proof: We prove and discuss only item 2 here. The proofs of the other results are left as exercises. Let $\{v_1, \ldots, v_k\}$ be an orthonormal basis for $\mathcal{S}$ and let $x \in \mathbb{R}^n$ be an arbitrary vector. Set

$$\begin{aligned} x_1 &= \sum_{i=1}^{k} (x^T v_i) v_i, \\ x_2 &= x - x_1. \end{aligned}$$

Then $x_1 \in \mathcal{S}$ and, since

$$\begin{aligned} x_2^T v_j &= x^T v_j - x_1^T v_j \\ &= x^T v_j - x^T v_j = 0, \end{aligned}$$

we see that x_2 is orthogonal to $v_1, \ldots, v_k$ and hence to any linear combination of these vectors. In other words, x_2 is orthogonal to any vector in $\mathcal{S}$. We have thus shown that $\mathcal{S} + \mathcal{S}^\perp = \mathbb{R}^n$. We also have that $\mathcal{S} \cap \mathcal{S}^\perp = 0$ since the only vector $s \in \mathcal{S}$ orthogonal to everything in $\mathcal{S}$ (i.e., including itself) is 0.

It is also easy to see directly that, when we have such direct sum decompositions, we can write vectors in a unique way with respect to the corresponding subspaces. Suppose, for example, that $x = x_1 + x_2 = x_1' + x_2'$, where $x_1, x_1' \in \mathcal{S}$ and $x_2, x_2' \in \mathcal{S}^\perp$. Then $(x_1' - x_1)^T (x_2' - x_2) = 0$ by definition of $\mathcal{S}^\perp$. But then $(x_1' - x_1)^T (x_1' - x_1) = 0$ since $x_2' - x_2 = -(x_1' - x_1)$ (which follows by rearranging the equation $x_1 + x_2 = x_1' + x_2'$). Thus, $x_1 = x_1'$ and $x_2 = x_2'$. $\square$

Theorem 3.12. *Let $A : \mathbb{R}^n \to \mathbb{R}^m$. Then*

1. $\mathcal{N}(A)^\perp = \mathcal{R}(A^T)$. *(Note: This holds only for finite-dimensional vector spaces.)*
2. $\mathcal{R}(A)^\perp = \mathcal{N}(A^T)$. *(Note: This also holds for infinite-dimensional vector spaces.)*

Proof: To prove the first part, take an arbitrary $x \in \mathcal{N}(A)$. Then $Ax = 0$ and this is equivalent to $y^T Ax = 0$ for all y. But $y^T Ax = (A^T y)^T x$. Thus, $Ax = 0$ if and only if x is orthogonal to all vectors of the form $A^T y$, i.e., $x \in \mathcal{R}(A^T)^\perp$. Since x was arbitrary, we have established that $\mathcal{N}(A)^\perp = \mathcal{R}(A^T)$.

The proof of the second part is similar and is left as an exercise. $\square$

Definition 3.13. *Let $A : \mathbb{R}^n \to \mathbb{R}^m$. Then $\{v \in \mathbb{R}^n : Av = 0\}$ is sometimes called the* **right nullspace** *of A. Similarly, $\{w \in \mathbb{R}^m : w^T A = 0\}$ is called the* **left nullspace** *of A. Clearly, the right nullspace is $\mathcal{N}(A)$ while the left nullspace is $\mathcal{N}(A^T)$.*

Theorem 3.12 and part 2 of Theorem 3.11 can be combined to give two very fundamental and useful decompositions of vectors in the domain and co-domain of a linear transformation A. See also Theorem 2.26.

Theorem 3.14 (Decomposition Theorem). *Let $A : \mathbb{R}^n \to \mathbb{R}^m$. Then*

1. *every vector v in the domain space $\mathbb{R}^n$ can be written in a unique way as $v = x + y$, where $x \in \mathcal{N}(A)$ and $y \in \mathcal{N}(A)^\perp = \mathcal{R}(A^T)$ (i.e., $\mathbb{R}^n = \mathcal{N}(\mathcal{A}) \oplus \mathcal{R}(A^T)$).*
2. *every vector w in the co-domain space $\mathbb{R}^m$ can be written in a unique way as $w = x+y$, where $x \in \mathcal{R}(A)$ and $y \in \mathcal{R}(A)^\perp = \mathcal{N}(A^T)$ (i.e., $\mathbb{R}^m = \mathcal{R}(\mathcal{A}) \oplus \mathcal{N}(A^T)$).*

This key theorem becomes very easy to remember by carefully studying and understanding Figure 3.1 in the next section.

3.5 Four Fundamental Subspaces

Consider a general matrix $A \in \mathbb{R}_r^{m \times n}$. When thought of as a linear transformation from $\mathbb{R}^n$ to $\mathbb{R}^m$, many properties of A can be developed in terms of the four fundamental subspaces

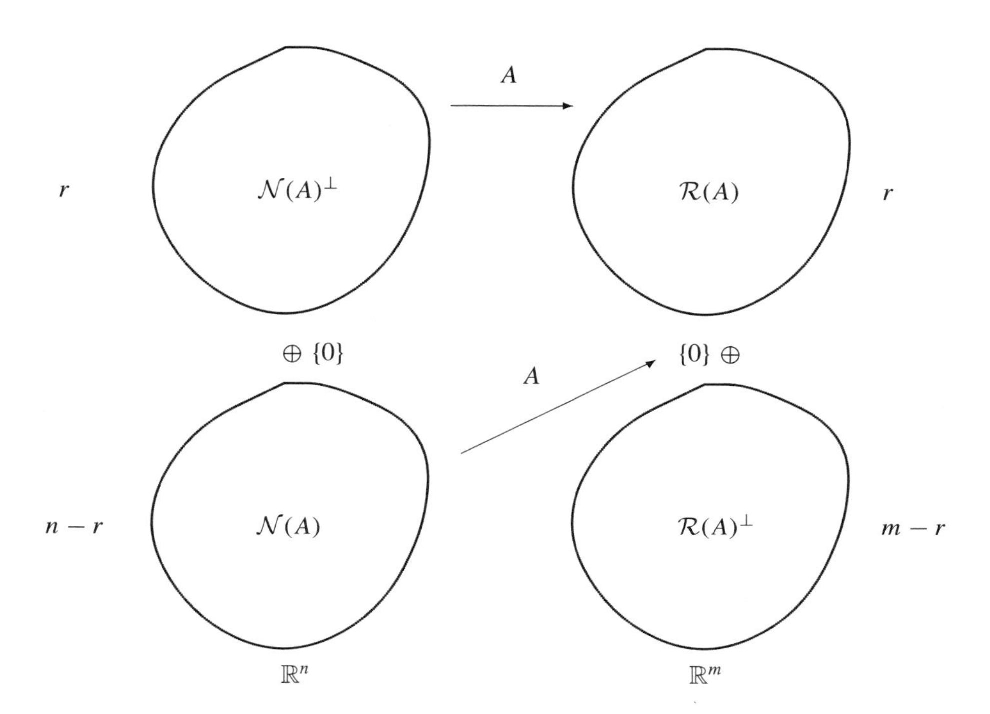

Figure 3.1. *Four fundamental subspaces.*

$\mathcal{R}(A)$, $\mathcal{R}(A)^{\perp}$, $\mathcal{N}(A)$, and $\mathcal{N}(A)^{\perp}$. Figure 3.1 makes many key properties seem almost obvious and we return to this figure frequently both in the context of linear transformations and in illustrating concepts such as controllability and observability.

Definition 3.15. *Let $\mathcal{V}$ and $\mathcal{W}$ be vector spaces and let $A : \mathcal{V} \rightarrow \mathcal{W}$ be a linear transformation.*

1. *A is* **onto** *(also called epic or surjective) if $\mathcal{R}(A) = \mathcal{W}$.*

2. *A is* **one-to-one** *or* **1–1** *(also called monic or injective) if $\mathcal{N}(A) = 0$. Two equivalent characterizations of A being 1–1 that are often easier to verify in practice are the following:*

   (a) $Av_1 = Av_2 \Longrightarrow v_1 = v_2$.

   (b) $v_1 \neq v_2 \Longrightarrow Av_1 \neq Av_2$.

Definition 3.16. *Let $A : \mathbb{R}^n \rightarrow \mathbb{R}^m$. Then* $\mathrm{rank}(A) = \dim \mathcal{R}(A)$. *This is sometimes called the* **column rank** *of A (maximum number of independent columns). The* **row rank** *of A is*

$\dim \mathcal{R}(A^T)$ *(maximum number of independent rows). The dual notion to rank is the* **nullity** *of A, sometimes denoted* $\text{nullity}(A)$ *or* $\text{corank}(A)$, *and is defined as* $\dim \mathcal{N}(A)$.

Theorem 3.17. *Let* $A : \mathbb{R}^n \to \mathbb{R}^m$. *Then* $\dim \mathcal{R}(A) = \dim \mathcal{N}(A)^\perp$. *(Note: Since* $\mathcal{N}(A)^\perp = \mathcal{R}(A^T)$, *this theorem is sometimes colloquially stated "row rank of A = column rank of A.")*

Proof: Define a linear transformation $T : \mathcal{N}(A)^\perp \to \mathcal{R}(A)$ by

$$Tv = Av \quad \text{for all } v \in \mathcal{N}(A)^\perp.$$

Clearly T is 1–1 (since $\mathcal{N}(T) = 0$). To see that T is also onto, take any $w \in \mathcal{R}(A)$. Then by definition there is a vector $x \in \mathbb{R}^n$ such that $Ax = w$. Write $x = x_1 + x_2$, where $x_1 \in \mathcal{N}(A)^\perp$ and $x_2 \in \mathcal{N}(A)$. Then $Ax_1 = w = Tx_1$ since $x_1 \in \mathcal{N}(A)^\perp$. The last equality shows that T is onto. We thus have that $\dim \mathcal{R}(A) = \dim \mathcal{N}(A)^\perp$ since it is easily shown that if $\{v_1, \ldots, v_r\}$ is a basis for $\mathcal{N}(A)^\perp$, then $\{Tv_1, \ldots, Tv_r\}$ is a basis for $\mathcal{R}(A)$. Finally, if we apply this and several previous results, the following string of equalities follows easily: "column rank of A" = $\text{rank}(A) = \dim \mathcal{R}(A) = \dim \mathcal{N}(A)^\perp = \dim \mathcal{R}(A^T) = \text{rank}(A^T)$ = "row rank of A." □

The following corollary is immediate. Like the theorem, it is a statement about equality of dimensions; the subspaces themselves are not necessarily in the same vector space.

Corollary 3.18. *Let* $A : \mathbb{R}^n \to \mathbb{R}^m$. *Then* $\dim \mathcal{N}(A) + \dim \mathcal{R}(A) = n$, *where n is the dimension of the domain of A.*

Proof: From Theorems 3.11 and 3.17 we see immediately that

$$\begin{aligned} n &= \dim \mathcal{N}(A) + \dim \mathcal{N}(A)^\perp \\ &= \dim \mathcal{N}(A) + \dim \mathcal{R}(A) . \end{aligned}$$ □

For completeness, we include here a few miscellaneous results about ranks of sums and products of matrices.

Theorem 3.19. *Let* $A, B \in \mathbb{R}^{n \times n}$. *Then*

1. $0 \le \text{rank}(A + B) \le \text{rank}(A) + \text{rank}(B)$.

2. $\text{rank}(A) + \text{rank}(B) - n \le \text{rank}(AB) \le \min\{\text{rank}(A),\ \text{rank}(B)\}$.

3. $\text{nullity}(B) \le \text{nullity}(AB) \le \text{nullity}(A) + \text{nullity}(B)$.

4. *if B is nonsingular,* $\text{rank}(AB) = \text{rank}(BA) = \text{rank}(A)$ *and* $\mathcal{N}(BA) = \mathcal{N}(A)$.

Part 4 of Theorem 3.19 suggests looking at the general problem of the four fundamental subspaces of matrix products. The basic results are contained in the following easily proved theorem.

Theorem 3.20. *Let $A \in \mathbb{R}^{m\times n}$, $B \in \mathbb{R}^{n\times p}$. Then*

1. $\mathcal{R}(AB) \subseteq \mathcal{R}(A)$.
2. $\mathcal{N}(AB) \supseteq \mathcal{N}(B)$.
3. $\mathcal{R}((AB)^T) \subseteq \mathcal{R}(B^T)$.
4. $\mathcal{N}((AB)^T) \supseteq \mathcal{N}(A^T)$.

The next theorem is closely related to Theorem 3.20 and is also easily proved. It is extremely useful in text that follows, especially when dealing with pseudoinverses and linear least squares problems.

Theorem 3.21. *Let $A \in \mathbb{R}^{m\times n}$. Then*

1. $\mathcal{R}(A) = \mathcal{R}(AA^T)$.
2. $\mathcal{R}(A^T) = \mathcal{R}(A^TA)$.
3. $\mathcal{N}(A) = \mathcal{N}(A^TA)$.
4. $\mathcal{N}(A^T) = \mathcal{N}(AA^T)$.

We now characterize 1–1 and onto transformations and provide characterizations in terms of rank and invertibility.

Theorem 3.22. *Let $A : \mathbb{R}^n \to \mathbb{R}^m$. Then*

1. *A is onto if and only if* $\operatorname{rank}(A) = m$ *(A has linearly independent rows or is said to have full row rank; equivalently, AA^T is nonsingular).*
2. *A is 1–1 if and only if* $\operatorname{rank}(A) = n$ *(A has linearly independent columns or is said to have full column rank; equivalently, A^TA is nonsingular).*

Proof: Proof of part 1: If A is onto, $\dim \mathcal{R}(A) = m = \operatorname{rank}(A)$. Conversely, let $y \in \mathbb{R}^m$ be arbitrary. Let $x = A^T(AA^T)^{-1}y \in \mathbb{R}^n$. Then $y = Ax$, i.e., $y \in \mathcal{R}(A)$, so A is onto.

Proof of part 2: If A is 1–1, then $\mathcal{N}(A) = 0$, which implies that $\dim \mathcal{N}(A)^{\perp} = n = \dim \mathcal{R}(A^T)$, and hence $\dim \mathcal{R}(A) = n$ by Theorem 3.17. Conversely, suppose $Ax_1 = Ax_2$. Then $A^TAx_1 = A^TAx_2$, which implies $x_1 = x_2$ since A^TA is invertible. Thus, A is 1–1. $\square$

Definition 3.23. *$A : \mathcal{V} \to \mathcal{W}$ is* **invertible** *(or bijective) if and only if it is 1–1 and onto. Note that if A is invertible, then* $\dim \mathcal{V} = \dim \mathcal{W}$. *Also, $A : \mathbb{R}^n \to \mathbb{R}^n$ is* **invertible** *or* **nonsingular** *if and only if* $\operatorname{rank}(A) = n$.

Note that in the special case when $A \in \mathbb{R}_n^{n\times n}$, the transformations A, A^T, and A^{-1} are all 1–1 and onto between the two spaces $\mathcal{N}(A)^{\perp}$ and $\mathcal{R}(A)$. The transformations A^T and A^{-1} have the same domain and range but are in general different maps unless A is orthogonal. Similar remarks apply to A and A^{-T}.

If a linear transformation is not invertible, it may still be right or left invertible. Definitions of these concepts are followed by a theorem characterizing left and right invertible transformations.

Definition 3.24. *Let $A : \mathcal{V} \rightarrow \mathcal{W}$. Then*

1. *A is said to be* **right invertible** *if there exists a right inverse transformation $A^{-R} : \mathcal{W} \rightarrow \mathcal{V}$ such that $AA^{-R} = I_w$, where I_w denotes the identity transformation on $\mathcal{W}$.*

2. *A is said to be* **left invertible** *if there exists a left inverse transformation $A^{-L} : \mathcal{W} \rightarrow \mathcal{V}$ such that $A^{-L}A = I_v$, where I_v denotes the identity transformation on $\mathcal{V}$.*

Theorem 3.25. *Let $A : \mathcal{V} \rightarrow \mathcal{W}$. Then*

1. *A is right invertible if and only if it is onto.*

2. *A is left invertible if and only if it is 1–1.*

Moreover, A is invertible if and only if it is both right and left invertible, i.e., both 1–1 and onto, in which case $A^{-1} = A^{-R} = A^{-L}$.

Note: From Theorem 3.22 we see that if $A : \mathbb{R}^n \rightarrow \mathbb{R}^m$ is onto, then a right inverse is given by $A^{-R} = A^T(AA^T)^{-1}$. Similarly, if A is 1–1, then a left inverse is given by $A^{-L} = (A^TA)^{-1}A^T$.

Theorem 3.26. *Let $A : \mathcal{V} \rightarrow \mathcal{V}$.*

1. *If there exists a unique right inverse A^{-R} such that $AA^{-R} = I$, then A is invertible.*

2. *If there exists a unique left inverse A^{-L} such that $A^{-L}A = I$, then A is invertible.*

Proof: We prove the first part and leave the proof of the second to the reader. Notice the following:

$$\begin{aligned} A(A^{-R} + A^{-R}A - I) &= AA^{-R} + AA^{-R}A - A \\ &= I + IA - A \quad \text{since } AA^{-R} = I \\ &= I. \end{aligned}$$

Thus, $(A^{-R} + A^{-R}A - I)$ must be a right inverse and, therefore, by uniqueness it must be the case that $A^{-R} + A^{-R}A - I = A^{-R}$. But this implies that $A^{-R}A = I$, i.e., that A^{-R} is a left inverse. It then follows from Theorem 3.25 that A is invertible. □

Example 3.27.

1. Let $A = [1 \;\; 2] : \mathbb{R}^2 \rightarrow \mathbb{R}^1$. Then A is onto. (*Proof:* Take any $\alpha \in \mathbb{R}^1$; then one can always find $v \in \mathbb{R}^2$ such that $[1 \;\; 2]\left[\begin{smallmatrix} v_1 \\ v_2 \end{smallmatrix}\right] = \alpha$). Obviously A has full row rank (=1) and $A^{-R} = \left[\begin{smallmatrix} 3 \\ -1 \end{smallmatrix}\right]$ is a right inverse. Also, it is clear that there are infinitely many right inverses for A. In Chapter 6 we characterize all right inverses of a matrix by characterizing all solutions of the linear matrix equation $AR = I$.

2. Let $A = \begin{bmatrix} 1 \\ 2 \end{bmatrix} : \mathbb{R}^1 \to \mathbb{R}^2$. Then A is 1–1. (*Proof*: The only solution to $0 = Av = \begin{bmatrix} 1 \\ 2 \end{bmatrix} v$ is $v = 0$, whence $\mathcal{N}(A) = 0$ so A is 1–1). It is now obvious that A has full column rank (=1) and $A^{-L} = [3 \quad -1]$ is a left inverse. Again, it is clear that there are infinitely many left inverses for A. In Chapter 6 we characterize all left inverses of a matrix by characterizing all solutions of the linear matrix equation $LA = I$.

3. The matrix

$$A = \begin{bmatrix} 1 & 1 & 1 \\ 2 & 1 & 1 \\ 3 & 1 & 1 \end{bmatrix} \in \mathbb{R}_2^{3\times 3},$$

when considered as a linear transformation on $\mathbb{R}^3$, is neither 1–1 nor onto. We give below bases for its four fundamental subspaces.

$$\mathcal{R}(A) = \mathrm{Sp}\left\{ \begin{bmatrix} 1 \\ 2 \\ 3 \end{bmatrix}, \begin{bmatrix} 1 \\ 1 \\ 1 \end{bmatrix} \right\}, \quad \mathcal{R}(A)^{\perp} = \mathrm{Sp}\left\{ \begin{bmatrix} 1 \\ -2 \\ 1 \end{bmatrix} \right\},$$

$$\mathcal{N}(A) = \mathrm{Sp}\left\{ \begin{bmatrix} 0 \\ 1 \\ -1 \end{bmatrix} \right\}, \quad \mathcal{N}(A)^{\perp} = \mathrm{Sp}\left\{ \begin{bmatrix} 1 \\ 0 \\ 0 \end{bmatrix}, \begin{bmatrix} 0 \\ 1 \\ 1 \end{bmatrix} \right\}.$$

EXERCISES

1. Let $A = \begin{bmatrix} 2 & 3 & 4 \\ 8 & 5 & 1 \end{bmatrix}$ and consider A as a linear transformation mapping $\mathbb{R}^3$ to $\mathbb{R}^2$. Find the matrix representation of A with respect to the bases

$$\left\{ \begin{bmatrix} 1 \\ 1 \\ 0 \end{bmatrix}, \begin{bmatrix} 0 \\ 1 \\ 1 \end{bmatrix}, \begin{bmatrix} 1 \\ 0 \\ 1 \end{bmatrix} \right\}$$

of $\mathbb{R}^3$ and

$$\left\{ \begin{bmatrix} 3 \\ 1 \end{bmatrix}, \begin{bmatrix} 2 \\ 1 \end{bmatrix} \right\}$$

of $\mathbb{R}^2$.

2. Consider the vector space $\mathbb{R}^{n\times n}$ over $\mathbb{R}$, let $\mathcal{S}$ denote the subspace of symmetric matrices, and let $\mathcal{R}$ denote the subspace of skew-symmetric matrices. For matrices $X, Y \in \mathbb{R}^{n\times n}$ define their inner product by $\langle X, Y\rangle = \mathrm{Tr}(X^T Y)$. Show that, with respect to this inner product, $\mathcal{R} = \mathcal{S}^{\perp}$.

3. Consider the differentiation operator $\mathcal{L}$ defined in Example 3.2.3. Is $\mathcal{L}$ 1–1? Is $\mathcal{L}$ onto?

4. Prove Theorem 3.4.

5. Prove Theorem 3.11.4.

6. Prove Theorem 3.12.2.

7. Determine bases for the four fundamental subspaces of the matrix

$$A = \begin{bmatrix} 1 & 1 & 1 & 0 \\ 1 & 2 & 2 & 1 \\ 2 & 5 & 5 & 3 \end{bmatrix}.$$

8. Suppose $A \in \mathbb{R}^{m \times n}$ has a left inverse. Show that A^T has a right inverse.

9. Let $A = \left[\begin{smallmatrix} 0 & 0 \\ 1 & 0 \end{smallmatrix}\right]$. Determine $\mathcal{N}(A)$ and $\mathcal{R}(A)$. Are they equal? Is this true in general? If this is true in general, prove it; if not, provide a counterexample.

10. Suppose $A \in \mathbb{R}_8^{19 \times 48}$. How many linearly independent solutions can be found to the homogeneous linear system $Ax = 0$?

11. Modify Figure 3.1 to illustrate the four fundamental subspaces associated with $A^T \in \mathbb{R}^{n \times m}$ thought of as a transformation from $\mathbb{R}^m$ to $\mathbb{R}^n$.

Chapter 4

Introduction to the Moore–Penrose Pseudoinverse

In this chapter we give a brief introduction to the Moore–Penrose pseudoinverse, a generalization of the inverse of a matrix. The Moore–Penrose pseudoinverse is defined for any matrix and, as is shown in the following text, brings great notational and conceptual clarity to the study of solutions to arbitrary systems of linear equations and linear least squares problems.

4.1 Definitions and Characterizations

Consider a linear transformation $A : \mathcal{X} \to \mathcal{Y}$, where $\mathcal{X}$ and $\mathcal{Y}$ are arbitrary finite-dimensional vector spaces. Define a transformation $T : \mathcal{N}(A)^{\perp} \to \mathcal{R}(A)$ by

$$Tx = Ax \quad \text{for all} \quad x \in \mathcal{N}(A)^{\perp}.$$

Then, as noted in the proof of Theorem 3.17, T is bijective (1–1 and onto), and hence we can define a unique inverse transformation $T^{-1} : \mathcal{R}(A) \to \mathcal{N}(A)^{\perp}$. This transformation can be used to give our first definition of A^{+}, the Moore–Penrose pseudoinverse of A. Unfortunately, the definition neither provides nor suggests a good computational strategy for determining A^{+}.

Definition 4.1. *With A and T as defined above, define a transformation $A^{+} : \mathcal{Y} \to \mathcal{X}$ by*

$$A^{+}y = T^{-1}y_1,$$

where $y = y_1 + y_2$ with $y_1 \in \mathcal{R}(A)$ and $y_2 \in \mathcal{R}(A)^{\perp}$. Then A^{+} is the **Moore–Penrose pseudoinverse** *of A.*

Although $\mathcal{X}$ and $\mathcal{Y}$ were arbitrary vector spaces above, let us henceforth consider the case $\mathcal{X} = \mathbb{R}^n$ and $\mathcal{Y} = \mathbb{R}^m$. We have thus defined A^{+} for all $A \in \mathbb{R}_r^{m \times n}$. A purely algebraic characterization of A^{+} is given in the next theorem, which was proved by Penrose in 1955; see [22].

Theorem 4.2. *Let $A \in \mathbb{R}_r^{m\times n}$. Then $G = A^+$ if and only if*

(P1) $AGA = A$.

(P2) $GAG = G$.

(P3) $(AG)^T = AG$.

(P4) $(GA)^T = GA$.

Furthermore, A^+ always exists and is unique.

Note that the inverse of a nonsingular matrix satisfies all four Penrose properties. Also, a right or left inverse satisfies no fewer than three of the four properties. Unfortunately, as with Definition 4.1, neither the statement of Theorem 4.2 nor its proof suggests a computational algorithm. However, the Penrose properties do offer the great virtue of providing a checkable criterion in the following sense. Given a matrix G that is a candidate for being the pseudoinverse of A, one need simply verify the four Penrose conditions (P1)–(P4). If G satisfies all four, then by uniqueness, it must be A^+. Such a verification is often relatively straightforward.

Example 4.3. Consider $A = \begin{bmatrix} 1 \\ 2 \end{bmatrix}$. Verify directly that $A^+ = \begin{bmatrix} \frac{1}{5} & \frac{2}{5} \end{bmatrix}$ satisfies (P1)–(P4). Note that other left inverses (for example, $A^{-L} = [3 \quad -1]$) satisfy properties (P1), (P2), and (P4) but not (P3).

Still another characterization of A^+ is given in the following theorem, whose proof can be found in [1, p. 19]. While not generally suitable for computer implementation, this characterization can be useful for hand calculation of small examples.

Theorem 4.4. *Let $A \in \mathbb{R}_r^{m\times n}$. Then*

$$A^+ = \lim_{\delta\to 0} (A^T A + \delta^2 I)^{-1} A^T \tag{4.1}$$

$$= \lim_{\delta\to 0} A^T (AA^T + \delta^2 I)^{-1}. \tag{4.2}$$

4.2 Examples

Each of the following can be derived or verified by using the above definitions or characterizations.

Example 4.5. $A^+ = A^T(AA^T)^{-1}$ if A is onto (independent rows) (A is right invertible).

Example 4.6. $A^+ = (A^T A)^{-1} A^T$ if A is 1–1 (independent columns) (A is left invertible).

Example 4.7. For any scalar α,

$$\alpha^+ = \begin{cases} \alpha^{-1} & \text{if } \alpha \neq 0, \\ 0 & \text{if } \alpha = 0. \end{cases}$$

Example 4.8. For any vector $v \in \mathbb{R}^n$,

$$v^+ = (v^T v)^+ v^T = \begin{cases} \frac{v^T}{v^T v} & \text{if } v \neq 0, \\ 0 & \text{if } v = 0. \end{cases}$$

Example 4.9. $\begin{bmatrix} 1 & 0 \\ 0 & 0 \end{bmatrix}^+ = \begin{bmatrix} 1 & 0 \\ 0 & 0 \end{bmatrix}.$

Example 4.10. $\begin{bmatrix} 1 & 1 \\ 1 & 1 \end{bmatrix}^+ = \begin{bmatrix} \frac{1}{4} & \frac{1}{4} \\ \frac{1}{4} & \frac{1}{4} \end{bmatrix}.$

4.3 Properties and Applications

This section presents some miscellaneous useful results on pseudoinverses. Many of these are used in the text that follows.

Theorem 4.11. *Let $A \in \mathbb{R}^{m \times n}$ and suppose $U \in \mathbb{R}^{m \times m}$, $V \in \mathbb{R}^{n \times n}$ are orthogonal (M is orthogonal if $M^T = M^{-1}$). Then*

$$(UAV)^+ = V^T A^+ U^T.$$

Proof: For the proof, simply verify that the expression above does indeed satisfy each of the four Penrose conditions. □

Theorem 4.12. *Let $S \in \mathbb{R}^{n \times n}$ be symmetric with $U^T S U = D$, where U is orthogonal and D is diagonal. Then $S^+ = U D^+ U^T$, where D^+ is again a diagonal matrix whose diagonal elements are determined according to Example 4.7.*

Theorem 4.13. *For all $A \in \mathbb{R}^{m \times n}$,*

1. $A^+ = (A^T A)^+ A^T = A^T (A A^T)^+$.
2. $(A^T)^+ = (A^+)^T$.

Proof: Both results can be proved using the limit characterization of Theorem 4.4. The proof of the first result is not particularly easy and does not even have the virtue of being especially illuminating. The interested reader can consult the proof in [1, p. 27]. The proof of the second result (which can also be proved easily by verifying the four Penrose conditions) is as follows:

$$\begin{aligned} (A^T)^+ &= \lim_{\delta \to 0} (AA^T + \delta^2 I)^{-1} A \\ &= \lim_{\delta \to 0} [A^T (AA^T + \delta^2 I)^{-1}]^T \\ &= [\lim_{\delta \to 0} A^T (AA^T + \delta^2 I)^{-1}]^T \\ &= (A^+)^T. \qquad \square \end{aligned}$$

Note that by combining Theorems 4.12 and 4.13 we can, in theory at least, compute the Moore–Penrose pseudoinverse of any matrix (since AA^T and A^TA are symmetric). This turns out to be a poor approach in finite-precision arithmetic, however (see, e.g., [7], [11], [23]), and better methods are suggested in text that follows.

Theorem 4.11 is suggestive of a "reverse-order" property for pseudoinverses of products of matrices such as exists for inverses of products. Unfortunately, in general,

$$(AB)^+ \neq B^+A^+ .$$

As an example consider $A = [0 \quad 1]$ and $B = \begin{bmatrix} 1 \\ 1 \end{bmatrix}$. Then

$$(AB)^+ = 1^+ = 1$$

while

$$B^+A^+ = \begin{bmatrix} \frac{1}{2} & \frac{1}{2} \end{bmatrix} \begin{bmatrix} 0 \\ 1 \end{bmatrix} = \frac{1}{2}.$$

However, necessary and sufficient conditions under which the reverse-order property does hold are known and we quote a couple of moderately useful results for reference.

Theorem 4.14. $(AB)^+ = B^+A^+$ *if and only if*

1. $\mathcal{R}(BB^TA^T) \subseteq \mathcal{R}(A^T)$

   and

2. $\mathcal{R}(A^TAB) \subseteq \mathcal{R}(B)$.

Proof: For the proof, see [9]. □

Theorem 4.15. $(AB)^+ = B_1^+A_1^+$, *where* $B_1 = A^+AB$ *and* $A_1 = AB_1B_1^+$.

Proof: For the proof, see [5]. □

Theorem 4.16. *If* $A \in \mathbb{R}_r^{n\times r}$, $B \in \mathbb{R}_r^{r\times m}$, *then* $(AB)^+ = B^+A^+$.

Proof: Since $A \in \mathbb{R}_r^{n\times r}$, then $A^+ = (A^TA)^{-1}A^T$, whence $A^+A = I_r$. Similarly, since $B \in \mathbb{R}_r^{r\times m}$, we have $B^+ = B^T(BB^T)^{-1}$, whence $BB^+ = I_r$. The result then follows by taking $B_1 = B$, $A_1 = A$ in Theorem 4.15. □

The following theorem gives some additional useful properties of pseudoinverses.

Theorem 4.17. *For all* $A \in \mathbb{R}^{m\times n}$,

1. $(A^+)^+ = A$.
2. $(A^TA)^+ = A^+(A^T)^+$, $(AA^T)^+ = (A^T)^+A^+$.
3. $\mathcal{R}(A^+) = \mathcal{R}(A^T) = \mathcal{R}(A^+A) = \mathcal{R}(A^TA)$.
4. $\mathcal{N}(A^+) = \mathcal{N}(AA^+) = \mathcal{N}((AA^T)^+) = \mathcal{N}(AA^T) = \mathcal{N}(A^T)$.
5. *If A is normal, then* $A^kA^+ = A^+A^k$ *and* $(A^k)^+ = (A^+)^k$ *for all integers* $k > 0$.

Note: Recall that $A \in \mathbb{R}^{n \times n}$ is *normal* if $AA^T = A^T A$. For example, if A is symmetric, skew-symmetric, or orthogonal, then it is normal. However, a matrix can be none of the preceding but still be normal, such as

$$A = \begin{bmatrix} a & b \\ -b & a \end{bmatrix}$$

for scalars $a, b \in \mathbb{R}$.

The next theorem is fundamental to facilitating a compact and unifying approach to studying the existence of solutions of (matrix) linear equations and linear least squares problems.

Theorem 4.18. *Suppose $A \in \mathbb{R}^{n \times p}$, $B \in \mathbb{R}^{n \times m}$. Then $\mathcal{R}(B) \subseteq \mathcal{R}(A)$ if and only if $AA^+B = B$.*

Proof: Suppose $\mathcal{R}(B) \subseteq \mathcal{R}(A)$ and take arbitrary $x \in \mathbb{R}^m$. Then $Bx \in \mathcal{R}(B) \subseteq \mathcal{R}(A)$, so there exists a vector $y \in \mathbb{R}^p$ such that $Ay = Bx$. Then we have

$$Bx = Ay = AA^+Ay = AA^+Bx,$$

where one of the Penrose properties is used above. Since x was arbitrary, we have shown that $B = AA^+B$.

To prove the converse, assume that $AA^+B = B$ and take arbitrary $y \in \mathcal{R}(B)$. Then there exists a vector $x \in \mathbb{R}^m$ such that $Bx = y$, whereupon

$$y = Bx = AA^+Bx \in \mathcal{R}(A). \qquad \square$$

EXERCISES

1. Use Theorem 4.4 to compute the pseudoinverse of $\begin{bmatrix} 1 & 1 \\ 2 & 2 \end{bmatrix}$.

2. If $x, y \in \mathbb{R}^n$, show that $(xy^T)^+ = (x^Tx)^+(y^Ty)^+yx^T$.

3. For $A \in \mathbb{R}^{m \times n}$, prove that $\mathcal{R}(A) = \mathcal{R}(AA^T)$ using only definitions and elementary properties of the Moore–Penrose pseudoinverse.

4. For $A \in \mathbb{R}^{m \times n}$, prove that $\mathcal{R}(A^+) = \mathcal{R}(A^T)$.

5. For $A \in \mathbb{R}^{p \times n}$ and $B \in \mathbb{R}^{m \times n}$, show that $\mathcal{N}(A) \subseteq \mathcal{N}(B)$ if and only if $BA^+A = B$.

6. Let $A \in \mathbb{R}^{n \times n}$, $B \in \mathbb{R}^{n \times m}$, and $D \in \mathbb{R}^{m \times m}$ and suppose further that D is nonsingular.

   (a) Prove or disprove that

   $$\begin{bmatrix} A & AB \\ 0 & D \end{bmatrix}^+ = \begin{bmatrix} A^+ & -A^+ABD^{-1} \\ 0 & D^{-1} \end{bmatrix}.$$

   (b) Prove or disprove that

   $$\begin{bmatrix} A & B \\ 0 & D \end{bmatrix}^+ = \begin{bmatrix} A^+ & -A^+BD^{-1} \\ 0 & D^{-1} \end{bmatrix}.$$

Chapter 5

Introduction to the Singular Value Decomposition

In this chapter we give a brief introduction to the singular value decomposition (SVD). We show that every matrix has an SVD and describe some useful properties and applications of this important matrix factorization. The SVD plays a key conceptual and computational role throughout (numerical) linear algebra and its applications.

5.1 The Fundamental Theorem

Theorem 5.1. *Let $A \in \mathbb{R}_r^{m \times n}$. Then there exist orthogonal matrices $U \in \mathbb{R}^{m \times m}$ and $V \in \mathbb{R}^{n \times n}$ such that*

$$A = U \Sigma V^T, \tag{5.1}$$

where $\Sigma = \begin{bmatrix} S & 0 \\ 0 & 0 \end{bmatrix}$, $S = \operatorname{diag}(\sigma_1, \ldots, \sigma_r) \in \mathbb{R}^{r \times r}$, and $\sigma_1 \geq \cdots \geq \sigma_r > 0$. More specifically, we have

$$A = [U_1 \;\; U_2] \begin{bmatrix} S & 0 \\ 0 & 0 \end{bmatrix} \begin{bmatrix} V_1^T \\ V_2^T \end{bmatrix} \tag{5.2}$$

$$= U_1 S V_1^T . \tag{5.3}$$

The submatrix sizes are all determined by r (which must be $\leq \min\{m, n\}$), i.e., $U_1 \in \mathbb{R}^{m \times r}$, $U_2 \in \mathbb{R}^{m \times (m-r)}$, $V_1 \in \mathbb{R}^{n \times r}$, $V_2 \in \mathbb{R}^{n \times (n-r)}$, and the 0-subblocks in Σ are compatibly dimensioned.

Proof: Since $A^T A \geq 0$ ($A^T A$ is symmetric and nonnegative definite; recall, for example, [24, Ch. 6]), its eigenvalues are all real and nonnegative. (Note: The rest of the proof follows analogously if we start with the observation that $AA^T \geq 0$ and the details are left to the reader as an exercise.) Denote the set of eigenvalues of $A^T A$ by $\{\sigma_i^2 \, , \; i \in \underline{n}\}$ with $\sigma_1 \geq \cdots \geq \sigma_r > 0 = \sigma_{r+1} = \cdots = \sigma_n$. Let $\{v_i \, , \; i \in \underline{n}\}$ be a set of corresponding orthonormal eigenvectors and let $V_1 = [v_1, \ldots, v_r]$, $V_2 = [v_{r+1}, \ldots, v_n]$. Letting $S = \operatorname{diag}(\sigma_1, \ldots, \sigma_r)$, we can write $A^T A V_1 = V_1 S^2$. Premultiplying by V_1^T gives $V_1^T A^T A V_1 = V_1^T V_1 S^2 = S^2$, the latter equality following from the orthonormality of the v_i vectors. Pre- and postmultiplying by S^{-1} gives the equation

$$S^{-1} V_1^T A^T A V_1 S^{-1} = I. \tag{5.4}$$

Turning now to the eigenvalue equations corresponding to the eigenvalues $\sigma_{r+1}, \ldots, \sigma_n$ we have that $A^T A V_2 = V_2 0 = 0$, whence $V_2^T A^T A V_2 = 0$. Thus, $A V_2 = 0$. Now define the matrix $U_1 \in \mathbb{R}^{m \times r}$ by $U_1 = A V_1 S^{-1}$. Then from (5.4) we see that $U_1^T U_1 = I$; i.e., the columns of U_1 are orthonormal. Choose any matrix $U_2 \in \mathbb{R}^{m \times (m-r)}$ such that $[U_1 \quad U_2]$ is orthogonal. Then

$$
\begin{aligned}
U^T A V &= \begin{bmatrix} U_1^T A V_1 & U_1^T A V_2 \\ U_2^T A V_1 & U_2^T A V_2 \end{bmatrix} \\
&= \begin{bmatrix} U_1^T A V_1 & 0 \\ U_2^T A V_1 & 0 \end{bmatrix}
\end{aligned}
$$

since $A V_2 = 0$. Referring to the equation $U_1 = A V_1 S^{-1}$ defining U_1, we see that $U_1^T A V_1 = S$ and $U_2^T A V_1 = U_2^T U_1 S = 0$. The latter equality follows from the orthogonality of the columns of U_1 and U_2. Thus, we see that, in fact, $U^T A V = \left[\begin{smallmatrix} S & 0 \\ 0 & 0 \end{smallmatrix}\right]$, and defining this matrix to be Σ completes the proof. □

Definition 5.2. *Let $A = U \Sigma V^T$ be an SVD of A as in Theorem 5.1.*

1. *The set $\{\sigma_1, \ldots, \sigma_r\}$ is called the set of (nonzero)* **singular values** *of the matrix A and is denoted $\Sigma(A)$. From the proof of Theorem 5.1 we see that $\sigma_i(A) = \lambda_i^{\frac{1}{2}}(A^T A) = \lambda_i^{\frac{1}{2}}(A A^T)$. Note that there are also* $\min\{m, n\} - r$ *zero singular values.*

2. *The columns of U are called the* **left singular vectors** *of A (and are the orthonormal eigenvectors of $A A^T$).*

3. *The columns of V are called the* **right singular vectors** *of A (and are the orthonormal eigenvectors of $A^T A$).*

Remark 5.3. The analogous complex case in which $A \in \mathbb{C}_r^{m \times n}$ is quite straightforward. The decomposition is $A = U \Sigma V^H$, where U and V are unitary and the proof is essentially identical, except for Hermitian transposes replacing transposes.

Remark 5.4. Note that U and V can be interpreted as changes of basis in both the domain and co-domain spaces with respect to which A then has a diagonal matrix representation. Specifically, let $\mathcal{L}$ denote A thought of as a linear transformation mapping $\mathbb{R}^n$ to $\mathbb{R}^m$. Then rewriting $A = U \Sigma V^T$ as $A V = U \Sigma$ we see that Mat $\mathcal{L}$ is Σ with respect to the bases $\{v_1, \ldots, v_n\}$ for $\mathbb{R}^n$ and $\{u_1, \ldots, u_m\}$ for $\mathbb{R}^m$ (see the discussion in Section 3.2). See also Remark 5.16.

Remark 5.5. The singular value decomposition is not unique. For example, an examination of the proof of Theorem 5.1 reveals that

- any orthonormal basis for $\mathcal{N}(A)$ can be used for V_2.
- there may be nonuniqueness associated with the columns of V_1 (and hence U_1) corresponding to multiple σ_i's.

- any U_2 can be used so long as $[U_1 \quad U_2]$ is orthogonal.
- columns of U and V can be changed (in tandem) by sign (or multiplier of the form $e^{j\theta}$ in the complex case).

What is unique, however, is the matrix Σ and the span of the columns of U_1, U_2, V_1, and V_2 (see Theorem 5.11). Note, too, that a "full SVD" (5.2) can always be constructed from a "compact SVD" (5.3).

Remark 5.6. Computing an SVD by working directly with the eigenproblem for $A^T A$ or AA^T is numerically poor in finite-precision arithmetic. Better algorithms exist that work directly on A via a sequence of orthogonal transformations; see, e.g., [7], [11], [25].

Example 5.7.

$$A = \begin{bmatrix} 1 & 0 \\ 0 & 1 \end{bmatrix} = U\ I\ U^T,$$

where U is an arbitrary 2×2 orthogonal matrix, is an SVD.

Example 5.8.

$$A = \begin{bmatrix} 1 & 0 \\ 0 & -1 \end{bmatrix} = \begin{bmatrix} \cos\theta & \sin\theta \\ -\sin\theta & \cos\theta \end{bmatrix} \begin{bmatrix} 1 & 0 \\ 0 & 1 \end{bmatrix} \begin{bmatrix} \cos\theta & \sin\theta \\ \sin\theta & -\cos\theta \end{bmatrix},$$

where θ is arbitrary, is an SVD.

Example 5.9.

$$A = \begin{bmatrix} 1 & 1 \\ 2 & 2 \\ 2 & 2 \end{bmatrix} = \begin{bmatrix} \frac{1}{3} & \frac{-2\sqrt{5}}{5} & \frac{2\sqrt{5}}{15} \\ \frac{2}{3} & \frac{\sqrt{5}}{5} & \frac{4\sqrt{5}}{15} \\ \frac{2}{3} & 0 & \frac{-\sqrt{5}}{3} \end{bmatrix} \begin{bmatrix} 3\sqrt{2} & 0 \\ 0 & 0 \\ 0 & 0 \end{bmatrix} \begin{bmatrix} \frac{\sqrt{2}}{2} & \frac{\sqrt{2}}{2} \\ \frac{\sqrt{2}}{2} & \frac{-\sqrt{2}}{2} \end{bmatrix}$$

$$= \begin{bmatrix} \frac{1}{3} \\ \frac{2}{3} \\ \frac{2}{3} \end{bmatrix} 3\sqrt{2} \begin{bmatrix} \frac{\sqrt{2}}{2} & \frac{\sqrt{2}}{2} \end{bmatrix}$$

is an SVD.

Example 5.10. Let $A \in \mathbb{R}^{n \times n}$ be symmetric and positive definite. Let V be an orthogonal matrix of eigenvectors that diagonalizes A, i.e., $V^T AV = \Lambda > 0$. Then $A = V \Lambda V^T$ is an SVD of A.

A factorization $U \Sigma V^T$ of an $m \times n$ matrix A qualifies as an SVD if U and V are orthogonal and Σ is an $m \times n$ "diagonal" matrix whose diagonal elements in the upper left corner are positive (and ordered). For example, if $A = U \Sigma V^T$ is an SVD of A, then $V \Sigma^T U^T$ is an SVD of A^T.

5.2 Some Basic Properties

Theorem 5.11. *Let $A \in \mathbb{R}^{m \times n}$ have a singular value decomposition $A = U \Sigma V^T$. Using the notation of Theorem 5.1, the following properties hold:*

1. $\operatorname{rank}(A) = r =$ *the number of nonzero singular values of A.*

2. *Let $U = [u_1, \ldots, u_m]$ and $V = [v_1, \ldots, v_n]$. Then A has the dyadic (or outer product) expansion*

$$A = \sum_{i=1}^{r} \sigma_i u_i v_i^T . \tag{5.5}$$

3. *The singular vectors satisfy the relations*

$$A v_i = \sigma_i u_i , \tag{5.6}$$
$$A^T u_i = \sigma_i v_i \tag{5.7}$$

for $i \in \underline{r}$.

4. *Let $U_1 = [u_1, \ldots, u_r]$, $U_2 = [u_{r+1}, \ldots, u_m]$, $V_1 = [v_1, \ldots, v_r]$, and $V_2 = [v_{r+1}, \ldots, v_n]$. Then*

   (a) $\mathcal{R}(U_1) = \mathcal{R}(A) = \mathcal{N}(A^T)^{\perp}$.

   (b) $\mathcal{R}(U_2) = \mathcal{R}(A)^{\perp} = \mathcal{N}(A^T)$.

   (c) $\mathcal{R}(V_1) = \mathcal{N}(A)^{\perp} = \mathcal{R}(A^T)$.

   (d) $\mathcal{R}(V_2) = \mathcal{N}(A) = \mathcal{R}(A^T)^{\perp}$.

Remark 5.12. Part 4 of the above theorem provides a numerically superior method for finding (orthonormal) bases for the four fundamental subspaces compared to methods based on, for example, reduction to row or column echelon form. Note that each subspace requires knowledge of the rank r. The relationship to the four fundamental subspaces is summarized nicely in Figure 5.1.

Remark 5.13. The elegance of the dyadic decomposition (5.5) as a sum of outer products and the key vector relations (5.6) and (5.7) explain why it is conventional to write the SVD as $A = U \Sigma V^T$ rather than, say, $A = U \Sigma V$.

Theorem 5.14. *Let $A \in \mathbb{R}^{m \times n}$ have a singular value decomposition $A = U \Sigma V^T$ as in Theorem 5.1. Then*

$$A^+ = V \Sigma^+ U^T , \tag{5.8}$$

where

$$\Sigma^+ = \begin{bmatrix} S^{-1} & 0 \\ 0 & 0 \end{bmatrix} \in \mathbb{R}^{n \times m}$$

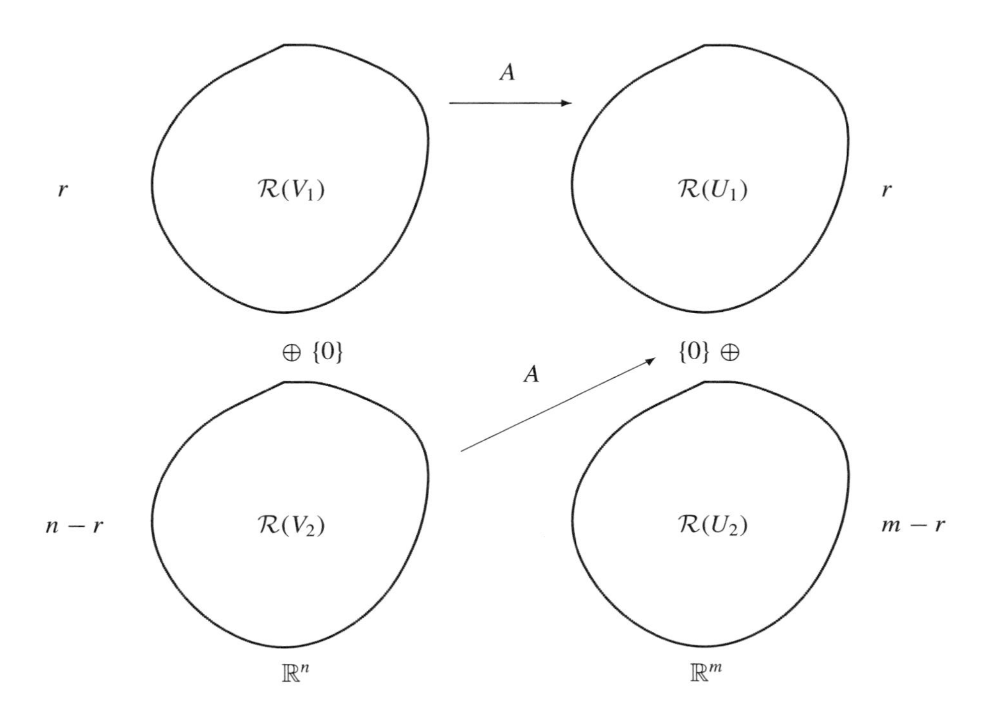

Figure 5.1. *SVD and the four fundamental subspaces.*

with the 0*-subblocks appropriately sized. Furthermore, if we let the columns of U and V be as defined in Theorem 5.11, then*

$$A^+ = V_1 S^{-1} U_1^T \tag{5.9}$$

$$= \sum_{i=1}^{r} \frac{1}{\sigma_i} v_i u_i^T . \tag{5.10}$$

Proof: The proof follows easily by verifying the four Penrose conditions. □

Remark 5.15. Note that none of the expressions above quite qualifies as an SVD of A^+ if we insist that the singular values be ordered from largest to smallest. However, a simple reordering accomplishes the task:

$$A^+ = \sum_{i=r}^{1} \frac{1}{\sigma_i} v_i u_i^T . \tag{5.11}$$

This can also be written in matrix terms by using the so-called reverse-order identity matrix (or exchange matrix) $P = [e_r, e_{r-1}, \ldots, e_2, e_1]$, which is clearly orthogonal and symmetric.

Then

$$A^+ = (V_1 P)(P S^{-1} P)(P U_1^T)$$

is the matrix version of (5.11). A "full SVD" can be similarly constructed.

Remark 5.16. Recall the linear transformation T used in the proof of Theorem 3.17 and in Definition 4.1. Since T is determined by its action on a basis, and since $\{v_1, \ldots, v_r\}$ is a basis for $\mathcal{N}(\mathcal{A})^{\perp}$, then T can be defined by $T v_i = \sigma_i u_i$, $i \in \underline{r}$. Similarly, since $\{u_1, \ldots, u_r\}$ is a basis for $\mathcal{R}(\mathcal{A})$, then T^{-1} can be defined by $T^{-1} u_i = \frac{1}{\sigma_i} v_i$, $i \in \underline{r}$. From Section 3.2, the matrix representation for T with respect to the bases $\{v_1, \ldots, v_r\}$ and $\{u_1, \ldots, u_r\}$ is clearly S, while the matrix representation for the inverse linear transformation T^{-1} with respect to the same bases is S^{-1}.

5.3 Row and Column Compressions

Row compression

Let $A \in \mathbb{R}^{m \times n}$ have an SVD given by (5.1). Then

$$\begin{aligned} U^T A &= \Sigma V^T \\ &= \begin{bmatrix} S & 0 \\ 0 & 0 \end{bmatrix} \begin{bmatrix} V_1^T \\ V_2^T \end{bmatrix} \\ &= \begin{bmatrix} S V_1^T \\ 0 \end{bmatrix} \in \mathbb{R}^{m \times n}. \end{aligned}$$

Notice that $\mathcal{N}(A) = \mathcal{N}(U^T A) = \mathcal{N}(S V_1^T)$ and the matrix $S V_1^T \in \mathbb{R}^{r \times n}$ has full row rank. In other words, premultiplication of A by U^T is an orthogonal transformation that "compresses" A by row transformations. Such a row compression can also be accomplished by orthogonal row transformations performed directly on A to reduce it to the form $\left[\begin{smallmatrix} R \\ 0 \end{smallmatrix}\right]$, where R is upper triangular. Both compressions are analogous to the so-called row-reduced echelon form which, when derived by a Gaussian elimination algorithm implemented in finite-precision arithmetic, is not generally as reliable a procedure.

Column compression

Again, let $A \in \mathbb{R}^{m \times n}$ have an SVD given by (5.1). Then

$$\begin{aligned} AV &= U \Sigma \\ &= [U_1 \quad U_2] \begin{bmatrix} S & 0 \\ 0 & 0 \end{bmatrix} \\ &= [U_1 S \quad 0] \in \mathbb{R}^{m \times n}. \end{aligned}$$

This time, notice that $\mathcal{R}(A) = \mathcal{R}(AV) = \mathcal{R}(U_1 S)$ and the matrix $U_1 S \in \mathbb{R}^{m \times r}$ has full column rank. In other words, postmultiplication of A by V is an orthogonal transformation that "compresses" A by column transformations. Such a compression is analogous to the so-called column-reduced echelon form, which is not generally a reliable procedure when performed by Gauss transformations in finite-precision arithmetic. For details, see, for example, [7], [11], [23], [25].

EXERCISES

1. Let $X \in \mathbb{R}^{m \times n}$. If $X^T X = 0$, show that $X = 0$.

2. Prove Theorem 5.1 starting from the observation that $AA^T \geq 0$.

3. Let $A \in \mathbb{R}^{n \times n}$ be symmetric but indefinite. Determine an SVD of A.

4. Let $x \in \mathbb{R}^m$, $y \in \mathbb{R}^n$ be nonzero vectors. Determine an SVD of the matrix $A \in \mathbb{R}_1^{m \times n}$ defined by $A = xy^T$.

5. Determine SVDs of the matrices

   (a) $\begin{bmatrix} -1 & 1 \\ 0 & -1 \end{bmatrix}$

   (b) $\begin{bmatrix} 1 & 0 \\ 1 & 0 \end{bmatrix}$.

6. Let $A \in \mathbb{R}^{m \times n}$ and suppose $W \in \mathbb{R}^{m \times m}$ and $Y \in \mathbb{R}^{n \times n}$ are orthogonal.

   (a) Show that A and WAY have the same singular values (and hence the same rank).

   (b) Suppose that W and Y are nonsingular but not necessarily orthogonal. Do A and WAY have the same singular values? Do they have the same rank?

7. Let $A \in \mathbb{R}_n^{n \times n}$. Use the SVD to determine a **polar factorization** of A, i.e., $A = QP$ where Q is orthogonal and $P = P^T > 0$. Note: this is analogous to the polar form $z = re^{i\theta}$ of a complex scalar z (where $i = j = \sqrt{-1}$).

Chapter 6

Linear Equations

In this chapter we examine existence and uniqueness of solutions of systems of linear equations. General linear systems of the form

$$AX = B\,; \quad A \in \mathbb{R}^{m\times n}, \; B \in \mathbb{R}^{m\times k}, \tag{6.1}$$

are studied and include, as a special case, the familiar vector system

$$Ax = b\,; \quad A \in \mathbb{R}^{n\times n}, \; b \in \mathbb{R}^{n}. \tag{6.2}$$

6.1 Vector Linear Equations

We begin with a review of some of the principal results associated with vector linear systems.

Theorem 6.1. *Consider the system of linear equations*

$$Ax = b\,; \quad A \in \mathbb{R}^{m\times n}, \; b \in \mathbb{R}^{m}. \tag{6.3}$$

1. *There exists a solution to (6.3) if and only if $b \in \mathcal{R}(A)$.*
2. *There exists a solution to (6.3) for all $b \in \mathbb{R}^m$ if and only if $\mathcal{R}(A) = \mathbb{R}^m$, i.e., A is onto; equivalently, there exists a solution if and only if* $\operatorname{rank}([A, b]) = \operatorname{rank}(A)$, *and this is possible only if $m \leq n$ (since $m = \dim \mathcal{R}(A) = \operatorname{rank}(A) \leq \min\{m, n\}$).*
3. *A solution to (6.3) is unique if and only if $\mathcal{N}(A) = 0$, i.e., A is 1–1.*
4. *There exists a unique solution to (6.3) for all $b \in \mathbb{R}^m$ if and only if A is nonsingular; equivalently, $A \in \mathbb{R}^{m\times m}$ and A has neither a 0 singular value nor a 0 eigenvalue.*
5. *There exists at most one solution to (6.3) for all $b \in \mathbb{R}^m$ if and only if the columns of A are linearly independent, i.e., $\mathcal{N}(A) = 0$, and this is possible only if $m \geq n$.*
6. *There exists a nontrivial solution to the homogeneous system $Ax = 0$ if and only if* $\operatorname{rank}(A) < n$.

Proof: The proofs are straightforward and can be consulted in standard texts on linear algebra. Note that some parts of the theorem follow directly from others. For example, to prove part 6, note that $x = 0$ is always a solution to the homogeneous system. Therefore, we must have the case of a nonunique solution, i.e., A is not 1–1, which implies $\operatorname{rank}(A) < n$ by part 3. □

6.2 Matrix Linear Equations

In this section we present some of the principal results concerning existence and uniqueness of solutions to the general matrix linear system (6.1). Note that the results of Theorem 6.1 follow from those below for the special case $k = 1$, while results for (6.2) follow by specializing even further to the case $m = n$.

Theorem 6.2 (Existence). *The matrix linear equation*

$$AX = B\ ;\quad A \in \mathbb{R}^{m\times n},\ \ B \in \mathbb{R}^{m\times k}, \tag{6.4}$$

has a solution if and only if $\mathcal{R}(B) \subseteq \mathcal{R}(A)$; equivalently, a solution exists if and only if $AA^+B = B$.

Proof: The subspace inclusion criterion follows essentially from the definition of the range of a matrix. The matrix criterion is Theorem 4.18. □

Theorem 6.3. *Let $A \in \mathbb{R}^{m\times n}$, $B \in \mathbb{R}^{m\times k}$ and suppose that $AA^+B = B$. Then any matrix of the form*

$$X = A^+B + (I - A^+A)Y, \quad \text{where } Y \in \mathbb{R}^{n\times k} \text{ is arbitrary}, \tag{6.5}$$

is a solution of

$$AX = B. \tag{6.6}$$

Furthermore, all solutions of (6.6) are of this form.

Proof: To verify that (6.5) is a solution, premultiply by A:

$$\begin{aligned} AX &= AA^+B + A(I - A^+A)Y \\ &= B + (A - AA^+A)Y \quad \text{by hypothesis} \\ &= B \quad \text{since } AA^+A = A \text{ by the first Penrose condition.} \end{aligned}$$

That all solutions are of this form can be seen as follows. Let Z be an arbitrary solution of (6.6), i.e., $AZ = B$. Then we can write

$$\begin{aligned} Z &\equiv A^+AZ + (I - A^+A)Z \\ &= A^+B + (I - A^+A)Z \end{aligned}$$

and this is clearly of the form (6.5). □

Remark 6.4. When A is square and nonsingular, $A^+ = A^{-1}$ and so $(I - A^+A) = 0$. Thus, there is no "arbitrary" component, leaving only the unique solution $X = A^{-1}B$.

Remark 6.5. It can be shown that the particular solution $X = A^+B$ is the solution of (6.6) that minimizes $\mathrm{Tr}X^TX$. ($\mathrm{Tr}(\cdot)$ denotes the trace of a matrix; recall that $\mathrm{Tr}X^TX = \sum_{i,j} x_{ij}^2$.)

Theorem 6.6 (Uniqueness). *A solution of the matrix linear equation*

$$AX = B; \quad A \in \mathbb{R}^{m\times n}, \; B \in \mathbb{R}^{m\times k} \tag{6.7}$$

is unique if and only if $A^+A = I$; equivalently, (6.7) has a unique solution if and only if $\mathcal{N}(A) = 0$.

Proof: The first equivalence is immediate from Theorem 6.3. The second follows by noting that $A^+A = I$ can occur only if $r = n$, where $r = \mathrm{rank}(A)$ (recall $r \le n$). But $\mathrm{rank}(A) = n$ if and only if A is 1–1 or $\mathcal{N}(A) = 0$. □

Example 6.7. Suppose $A \in \mathbb{R}^{n\times n}$. Find all solutions of the homogeneous system $Ax = 0$.
Solution:

$$\begin{aligned} x &= A^+0 + (I - A^+A)y \\ &= (I - A^+A)y, \end{aligned}$$

where $y \in \mathbb{R}^n$ is arbitrary. Hence, there exists a nonzero solution if and only if $A^+A \ne I$. This is equivalent to either $\mathrm{rank}(A) = r < n$ or A being singular. Clearly, if there exists a nonzero solution, it is not unique.

Computation: Since y is arbitrary, it is easy to see that all solutions are generated from a basis for $\mathcal{R}(I - A^+A)$. But if A has an SVD given by $A = U\Sigma V^T$, then it is easily checked that $I - A^+A = V_2V_2^T$ and $\mathcal{R}(V_2V_2^T) = \mathcal{R}(V_2) = \mathcal{N}(A)$.

Example 6.8. Characterize all right inverses of a matrix $A \in \mathbb{R}^{m\times n}$; equivalently, find all solutions R of the equation $AR = I_m$. Here, we write I_m to emphasize the $m \times m$ identity matrix.

Solution: There exists a right inverse if and only if $\mathcal{R}(I_m) \subseteq \mathcal{R}(A)$ and this is equivalent to $AA^+I_m = I_m$. Clearly, this can occur if and only if $\mathrm{rank}(A) = r = m$ (since $r \le m$) and this is equivalent to A being onto (A^+ is then a right inverse). All right inverses of A are then of the form

$$\begin{aligned} R &= A^+I_m + (I_n - A^+A)Y \\ &= A^+ + (I - A^+A)Y, \end{aligned}$$

where $Y \in \mathbb{R}^{n\times m}$ is arbitrary. There is a unique right inverse if and only if $A^+A = I$ ($\mathcal{N}(A) = 0$), in which case A must be invertible and $R = A^{-1}$.

Example 6.9. Consider the system of linear first-order difference equations

$$x_{k+1} = Ax_k + Bu_k; \quad x_0 \text{ given}; \; k \ge 0, \tag{6.8}$$

with $A \in \mathbb{R}^{n \times n}$ and $B \in \mathbb{R}^{n \times m}$ ($n \geq 1$, $m \geq 1$). The vector x_k in linear system theory is known as the **state vector** at time k while u_k is the **input (control) vector**. The general solution of (6.8) is given by

$$x_k = A^k x_0 + \sum_{j=0}^{k-1} A^{k-1-j} B u_j \tag{6.9}$$

$$= A^k x_0 + \left[B, AB, \ldots, A^{k-1}B\right] \begin{bmatrix} u_{k-1} \\ u_{k-2} \\ \vdots \\ u_0 \end{bmatrix} \tag{6.10}$$

for $k \geq 1$. We might now ask the question: Given $x_0 = 0$, does there exist an input sequence $\{u_j\}_{j=0}^{k-1}$ such that x_k takes an arbitrary value in $\mathbb{R}^n$? In linear system theory, this is a question of **reachability**. Since $m \geq 1$, from the fundamental Existence Theorem, Theorem 6.2, we see that (6.8) is reachable if and only if

$$\mathcal{R}(\left[B, AB, \ldots, A^{n-1}B\right]) = \mathbb{R}^n$$

or, equivalently, if and only if

$$\operatorname{rank}\left[B, AB, \ldots, A^{n-1}B\right] = n.$$

A related question is the following: Given an arbitrary initial vector x_0, does there exist an input sequence $\{u_j\}_{j=0}^{n-1}$ such that $x_n = 0$? In linear system theory, this is called **controllability**. Again from Theorem 6.2, we see that (6.8) is controllable if and only if

$$\mathcal{R}(A^n) \subseteq \mathcal{R}(\left[B, AB, \ldots, A^{n-1}B\right]).$$

Clearly, reachability always implies controllability and, if A is nonsingular, controllability and reachability are equivalent. The matrices $A = \left[\begin{smallmatrix} 0 & 1 \\ 0 & 0 \end{smallmatrix}\right]$ and $B = \left[\begin{smallmatrix} 1 \\ 0 \end{smallmatrix}\right]$ provide an example of a system that is controllable but not reachable.

The above are standard conditions with analogues for continuous-time models (i.e., linear differential equations). There are many other algebraically equivalent conditions.

Example 6.10. We now introduce an **output vector** y_k to the system (6.8) of Example 6.9 by appending the equation

$$y_k = Cx_k + Du_k \tag{6.11}$$

with $C \in \mathbb{R}^{p \times n}$ and $D \in \mathbb{R}^{p \times m}$ ($p \geq 1$). We can then pose some new questions about the overall system that are dual in the system-theoretic sense to reachability and controllability. The answers are cast in terms that are dual in the linear algebra sense as well. The condition dual to reachability is called **observability**: When does knowledge of $\{u_j\}_{j=0}^{n-1}$ and $\{y_j\}_{j=0}^{n-1}$ suffice to determine (uniquely) x_0? As a dual to controllability, we have the notion of **reconstructibility**: When does knowledge of $\{u_j\}_{j=0}^{n-1}$ and $\{y_j\}_{j=0}^{n-1}$ suffice to determine (uniquely) x_n? The fundamental duality result from linear system theory is the following:

(A, B) *is reachable* [*controllable*] *if and only if* (A^T, B^T) *is observable* [*reconstructible*].

To derive a condition for observability, notice that

$$y_k = CA^k x_0 + \sum_{j=0}^{k-1} CA^{k-1-j} Bu_j + Du_k. \tag{6.12}$$

Thus,

$$\begin{bmatrix} y_0 - Du_0 \\ y_1 - CBu_0 - Du_1 \\ \vdots \\ y_{n-1} - \sum_{j=0}^{n-2} CA^{n-2-j} Bu_j - Du_{n-1} \end{bmatrix} = \begin{bmatrix} C \\ CA \\ \vdots \\ CA^{n-1} \end{bmatrix} x_0. \tag{6.13}$$

Let v denote the (known) vector on the left-hand side of (6.13) and let R denote the matrix on the right-hand side. Then, by definition, $v \in \mathcal{R}(R)$, so a solution exists. By the fundamental Uniqueness Theorem, Theorem 6.6, the solution is then unique if and only if $\mathcal{N}(R) = 0$, or, equivalently, if and only if

$$\operatorname{rank} \begin{bmatrix} C \\ CA \\ \vdots \\ CA^{n-1} \end{bmatrix} = n.$$

6.3 A More General Matrix Linear Equation

Theorem 6.11. *Let $A \in \mathbb{R}^{m \times n}$, $B \in \mathbb{R}^{m \times q}$, and $C \in \mathbb{R}^{p \times q}$. Then the equation*

$$AXC = B \tag{6.14}$$

has a solution if and only if $AA^+BC^+C = B$, in which case the general solution is of the form

$$X = A^+BC^+ + Y - A^+AYCC^+, \tag{6.15}$$

where $Y \in \mathbb{R}^{n \times p}$ is arbitrary.

A compact matrix criterion for uniqueness of solutions to (6.14) requires the notion of the Kronecker product of matrices for its statement. Such a criterion ($CC^+ \otimes A^+A = I$) is stated and proved in Theorem 13.27.

6.4 Some Useful and Interesting Inverses

In many applications, the coefficient matrices of interest are square and nonsingular. Listed below is a small collection of useful matrix identities, particularly for block matrices, associated with matrix inverses. In these identities, $A \in \mathbb{R}^{n \times n}$, $B \in \mathbb{R}^{n \times m}$, $C \in \mathbb{R}^{m \times n}$, and $D \in \mathbb{R}^{m \times m}$. Invertibility is assumed for any component or subblock whose inverse is indicated. Verification of each identity is recommended as an exercise for the reader.

1. $(A + BDC)^{-1} = A^{-1} - A^{-1}B(D^{-1} + CA^{-1}B)^{-1}CA^{-1}$.
   This result is known as the **Sherman–Morrison–Woodbury formula**. It has many applications (and is frequently "rediscovered") including, for example, formulas for the inverse of a sum of matrices such as $(A + D)^{-1}$ or $(A^{-1} + D^{-1})^{-1}$. It also yields very efficient "updating" or "downdating" formulas in expressions such as $(A + xx^T)^{-1}$ (with symmetric $A \in \mathbb{R}^{n\times n}$ and $x \in \mathbb{R}^n$) that arise in optimization theory.

2. $\begin{bmatrix} I & B \\ 0 & I \end{bmatrix}^{-1} = \begin{bmatrix} I & -B \\ 0 & I \end{bmatrix}; \begin{bmatrix} I & 0 \\ C & I \end{bmatrix}^{-1} = \begin{bmatrix} I & 0 \\ -C & I \end{bmatrix}$.

3. $\begin{bmatrix} I & B \\ 0 & -I \end{bmatrix}^{-1} = \begin{bmatrix} I & B \\ 0 & -I \end{bmatrix}; \begin{bmatrix} I & 0 \\ C & -I \end{bmatrix}^{-1} = \begin{bmatrix} I & 0 \\ C & -I \end{bmatrix}$.
   Both of these matrices satisfy the matrix equation $X^2 = I$ from which it is obvious that $X^{-1} = X$. Note that the positions of the I and $-I$ blocks may be exchanged.

4. $\begin{bmatrix} A & B \\ 0 & D \end{bmatrix}^{-1} = \begin{bmatrix} A^{-1} & -A^{-1}BD^{-1} \\ 0 & D^{-1} \end{bmatrix}$.

5. $\begin{bmatrix} A & 0 \\ C & D \end{bmatrix}^{-1} = \begin{bmatrix} A^{-1} & 0 \\ -D^{-1}CA^{-1} & D^{-1} \end{bmatrix}$.

6. $\begin{bmatrix} I + BC & B \\ C & I \end{bmatrix}^{-1} = \begin{bmatrix} I & -B \\ -C & I + CB \end{bmatrix}$.

7. $\begin{bmatrix} A & B \\ C & D \end{bmatrix}^{-1} = \begin{bmatrix} A^{-1} + A^{-1}BECA^{-1} & -A^{-1}BE \\ -ECA^{-1} & E \end{bmatrix}$,
   where $E = (D - CA^{-1}B)^{-1}$ (E is the inverse of the Schur complement of A). This result follows easily from the block LU factorization in property 16 of Section 1.4.

8. $\begin{bmatrix} A & B \\ C & D \end{bmatrix}^{-1} = \begin{bmatrix} F & -FBD^{-1} \\ -D^{-1}CF & D^{-1} + D^{-1}CFBD^{-1} \end{bmatrix}$,
   where $F = (A - BD^{-1}C)^{-1}$. This result follows easily from the block UL factorization in property 17 of Section 1.4.

EXERCISES

1. As in Example 6.8, characterize all left inverses of a matrix $A \in \mathbb{R}^{m\times n}$.

2. Let $A \in \mathbb{R}^{m\times n}$, $B \in \mathbb{R}^{m\times k}$ and suppose A has an SVD as in Theorem 5.1. Assuming $\mathcal{R}(B) \subseteq \mathcal{R}(A)$, characterize all solutions of the matrix linear equation

$$AX = B$$

   in terms of the SVD of A.

3. Let $x, y \in \mathbb{R}^n$ and suppose further that $x^T y \neq 1$. Show that

$$(I - xy^T)^{-1} = I - \frac{1}{x^T y - 1} xy^T.$$

4. Let $x, y \in \mathbb{R}^n$ and suppose further that $x^T y \neq 1$. Show that

$$\begin{bmatrix} I & x \\ y^T & 1 \end{bmatrix}^{-1} = \begin{bmatrix} I + cxy^T & -cx \\ -cy^T & c \end{bmatrix},$$

where $c = 1/(1 - x^T y)$.

5. Let $A \in \mathbb{R}_n^{n \times n}$ and let A^{-1} have columns $c_1, \ldots, c_n$ and individual elements γ_{ij}. Assume that $\gamma_{ji} \neq 0$ for some i and j. Show that the matrix $B = A - \frac{1}{\gamma_{ji}} e_i e_j^T$ (i.e., A with $\frac{1}{\gamma_{ji}}$ subtracted from its (ij)th element) is singular.
Hint: Show that $c_i \in \mathcal{N}(B)$.

6. As in Example 6.10, check directly that the condition for reconstructibility takes the form

$$\mathcal{N} \begin{bmatrix} C \\ CA \\ \vdots \\ CA^{n-1} \end{bmatrix} \subseteq \mathcal{N}(A^n).$$

Chapter 7

Projections, Inner Product Spaces, and Norms

7.1 Projections

Definition 7.1. *Let $\mathcal{V}$ be a vector space with $\mathcal{V} = \mathcal{X} \oplus \mathcal{Y}$. By Theorem 2.26, every $v \in \mathcal{V}$ has a unique decomposition $v = x + y$ with $x \in \mathcal{X}$ and $y \in \mathcal{Y}$. Define $P_{\mathcal{X},\mathcal{Y}} : \mathcal{V} \to \mathcal{X} \subseteq \mathcal{V}$ by*

$$P_{\mathcal{X},\mathcal{Y}} v = x \quad \textit{for all } v \in \mathcal{V}.$$

$P_{\mathcal{X},\mathcal{Y}}$ is called the **(oblique) projection on $\mathcal{X}$ along $\mathcal{Y}$***.*

Figure 7.1 displays the projection of v on both $\mathcal{X}$ and $\mathcal{Y}$ in the case $\mathcal{V} = \mathbb{R}^2$.

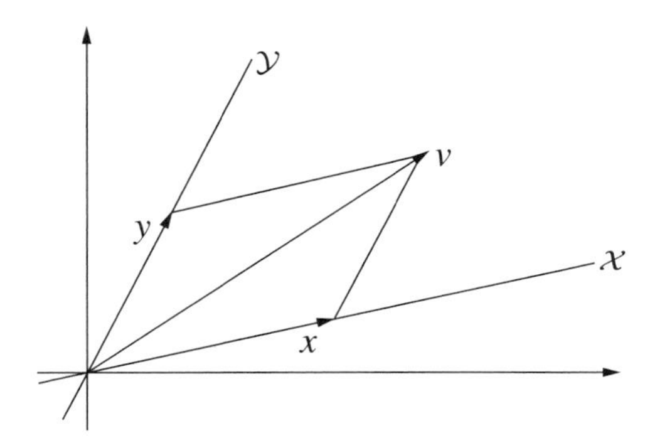

Figure 7.1. *Oblique projections.*

Theorem 7.2. *$P_{\mathcal{X},\mathcal{Y}}$ is linear and $P^2_{\mathcal{X},\mathcal{Y}} = P_{\mathcal{X},\mathcal{Y}}$.*

Theorem 7.3. *A linear transformation P is a projection if and only if it is* **idempotent***, i.e., $P^2 = P$. Also, P is a projection if and only if $I - P$ is a projection. In fact, $P_{\mathcal{Y},\mathcal{X}} = I - P_{\mathcal{X},\mathcal{Y}}$.*

Proof: Suppose P is a projection, say on $\mathcal{X}$ along $\mathcal{Y}$ (using the notation of Definition 7.1).

Let $v \in \mathcal{V}$ be arbitrary. Then $Pv = P(x+y) = Px = x$. Moreover, $P^2v = PPv = Px = x = Pv$. Thus, $P^2 = P$. Conversely, suppose $P^2 = P$. Let $\mathcal{X} = \{v \in \mathcal{V} : Pv = v\}$ and $\mathcal{Y} = \{v \in \mathcal{V} : Pv = 0\}$. It is easy to check that $\mathcal{X}$ and $\mathcal{Y}$ are subspaces. We now prove that $\mathcal{V} = \mathcal{X} \oplus \mathcal{Y}$. First note that if $v \in \mathcal{X}$, then $Pv = v$. If $v \in \mathcal{Y}$, then $Pv = 0$. Hence if $v \in \mathcal{X} \cap \mathcal{Y}$, then $v = 0$. Now let $v \in \mathcal{V}$ be arbitrary. Then $v = Pv + (I-P)v$. Let $x = Pv$, $y = (I-P)v$. Then $Px = P^2v = Pv = x$ so $x \in \mathcal{X}$, while $Py = P(I-P)v = Pv - P^2v = 0$ so $y \in \mathcal{Y}$. Thus, $\mathcal{V} = \mathcal{X} \oplus \mathcal{Y}$ and the projection on $\mathcal{X}$ along $\mathcal{Y}$ is P. Essentially the same argument shows that $I - P$ is the projection on $\mathcal{Y}$ along $\mathcal{X}$. □

Definition 7.4. *In the special case where $\mathcal{Y} = \mathcal{X}^\perp$, $P_{\mathcal{X},\mathcal{X}^\perp}$ is called an* **orthogonal projection** *and we then use the notation $P_{\mathcal{X}} = P_{\mathcal{X},\mathcal{X}^\perp}$.*

Theorem 7.5. *$P \in \mathbb{R}^{n\times n}$ is the matrix of an orthogonal projection (onto $\mathcal{R}(P)$) if and only if $P^2 = P = P^T$.*

Proof: Let P be an orthogonal projection (on $\mathcal{X}$, say, along $\mathcal{X}^\perp$) and let $x, y \in \mathbb{R}^n$ be arbitrary. Note that $(I-P)x = (I - P_{\mathcal{X},\mathcal{X}^\perp})x = P_{\mathcal{X}^\perp,\mathcal{X}}x$ by Theorem 7.3. Thus, $(I-P)x \in \mathcal{X}^\perp$. Since $Py \in \mathcal{X}$, we have $(Py)^T(I-P)x = y^TP^T(I-P)x = 0$. Since x and y were arbitrary, we must have $P^T(I-P) = 0$. Hence $P^T = P^TP = P$, with the second equality following since P^TP is symmetric. Conversely, suppose P is a symmetric projection matrix and let x be arbitrary. Write $x = Px + (I-P)x$. Then $x^TP^T(I-P)x = x^TP(I-P)x = 0$. Thus, since $Px \in \mathcal{R}(P)$, then $(I-P)x \in \mathcal{R}(P)^\perp$ and P must be an orthogonal projection. □

7.1.1 The four fundamental orthogonal projections

Using the notation of Theorems 5.1 and 5.11, let $A \in \mathbb{R}^{m\times n}$ with SVD $A = U\Sigma V^T = U_1SV_1^T$. Then

$$
\begin{aligned}
P_{\mathcal{R}(A)} &= AA^+ &&= U_1U_1^T &&= \sum_{i=1}^{r} u_iu_i^T,\\
P_{\mathcal{R}(A)^\perp} &= I - AA^+ &&= U_2U_2^T &&= \sum_{i=r+1}^{m} u_iu_i^T,\\
P_{\mathcal{N}(A)} &= I - A^+A &&= V_2V_2^T &&= \sum_{i=r+1}^{n} v_iv_i^T,\\
P_{\mathcal{N}(A)^\perp} &= A^+A &&= V_1V_1^T &&= \sum_{i=1}^{r} v_iv_i^T
\end{aligned}
$$

are easily checked to be (unique) orthogonal projections onto the respective four fundamental subspaces.

Example 7.6. Determine the orthogonal projection of a vector $v \in \mathbb{R}^n$ on another nonzero vector $w \in \mathbb{R}^n$.

Solution: Think of the vector w as an element of the one-dimensional subspace $\mathcal{R}(w)$. Then the desired projection is simply

$$\begin{aligned} P_{\mathcal{R}(w)}v &= ww^+v \\ &= \frac{ww^Tv}{w^Tw} \quad \text{(using Example 4.8)} \\ &= \left(\frac{w^Tv}{w^Tw}\right)w. \end{aligned}$$

Moreover, the vector z that is orthogonal to w and such that $v = Pv + z$ is given by $z = P_{\mathcal{R}(w)^\perp}v = (I - P_{\mathcal{R}(w)})v = v - \left(\frac{w^Tv}{w^Tw}\right)w$. See Figure 7.2. A direct calculation shows that z and w are, in fact, orthogonal:

$$w^Tz = w^Tv - \left(\frac{w^Tv}{w^Tw}\right)w^Tw = w^Tv - w^Tv = 0.$$

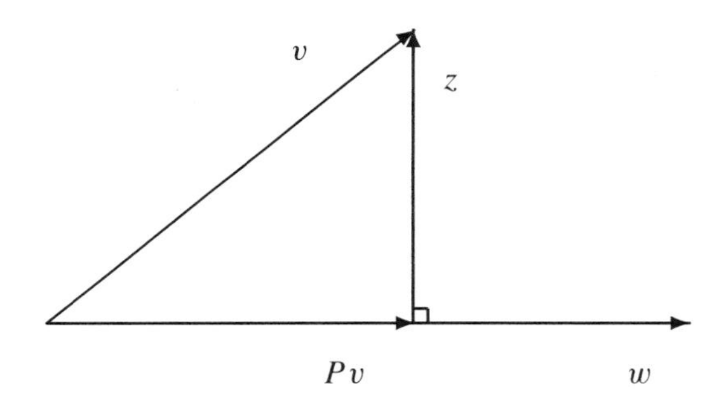

Figure 7.2. *Orthogonal projection on a "line."*

Example 7.7. Recall the proof of Theorem 3.11. There, $\{v_1, \ldots, v_k\}$ was an orthornormal basis for a subset $\mathcal{S}$ of $\mathbb{R}^n$. An arbitrary vector $x \in \mathbb{R}^n$ was chosen and a formula for x_1 appeared rather mysteriously. The expression for x_1 is simply the orthogonal projection of x on $\mathcal{S}$. Specifically,

$$P_{\mathcal{S}}x = VV^+x = VV^Tx = \left(\sum_{i=1}^{k} v_iv_i^T\right)x = \sum_{i=1}^{k}(v_i^Tx)v_i = x_1.$$

Example 7.8. Recall the diagram of the four fundamental subspaces. The indicated direct sum decompositions of the domain $\mathbb{R}^n$ and co-domain $\mathbb{R}^m$ are given easily as follows.

Let $x \in \mathbb{R}^n$ be an arbitrary vector. Then

$$\begin{aligned} x &= P_{\mathcal{N}(A)^\perp}x + P_{\mathcal{N}(A)}x \\ &= A^+Ax + (I - A^+A)x \\ &= V_1V_1^Tx + V_2V_2^Tx \quad (\text{recall } VV^T = I). \end{aligned}$$

Similarly, let $y \in \mathbb{R}^m$ be an arbitrary vector. Then

$$\begin{aligned} y &= P_{\mathcal{R}(A)}y + P_{\mathcal{R}(A)^\perp}y \\ &= AA^+y + (I - AA^+)y \\ &= U_1U_1^Ty + U_2U_2^Ty \quad (\text{recall } UU^T = I). \end{aligned}$$

Example 7.9. Let

$$A = \begin{bmatrix} 1 & 1 & 0 \\ 1 & 1 & 0 \end{bmatrix}.$$

Then

$$A^+ = \begin{bmatrix} 1/4 & 1/4 \\ 1/4 & 1/4 \\ 0 & 0 \end{bmatrix}$$

and we can decompose the vector $[2 \;\; 3 \;\; 4]^T$ uniquely into the sum of a vector in $\mathcal{N}(A)^\perp$ and a vector in $\mathcal{N}(A)$, respectively, as follows:

$$\begin{aligned} \begin{bmatrix} 2 \\ 3 \\ 4 \end{bmatrix} &= A^+Ax + (I - A^+A)x \\ &= \begin{bmatrix} 1/2 & 1/2 & 0 \\ 1/2 & 1/2 & 0 \\ 0 & 0 & 0 \end{bmatrix}\begin{bmatrix} 2 \\ 3 \\ 4 \end{bmatrix} + \begin{bmatrix} 1/2 & -1/2 & 0 \\ -1/2 & 1/2 & 0 \\ 0 & 0 & 1 \end{bmatrix}\begin{bmatrix} 2 \\ 3 \\ 4 \end{bmatrix} \\ &= \begin{bmatrix} 5/2 \\ 5/2 \\ 0 \end{bmatrix} + \begin{bmatrix} -1/2 \\ 1/2 \\ 4 \end{bmatrix}. \end{aligned}$$

7.2 Inner Product Spaces

Definition 7.10. *Let $\mathcal{V}$ be a vector space over $\mathbb{R}$. Then $\langle \cdot, \cdot \rangle : \mathcal{V} \times \mathcal{V} \to \mathbb{R}$ is a* **real inner product** *if*

1. *$\langle x, x \rangle \geq 0$ for all $x \in \mathcal{V}$ and $\langle x, x \rangle = 0$ if and only if $x = 0$.*
2. *$\langle x, y \rangle = \langle y, x \rangle$ for all $x, y \in \mathcal{V}$.*
3. *$\langle x, \alpha y_1 + \beta y_2 \rangle = \alpha \langle x, y_1 \rangle + \beta \langle x, y_2 \rangle$ for all $x, y_1, y_2 \in \mathcal{V}$ and for all $\alpha, \beta \in \mathbb{R}$.*

Example 7.11. Let $\mathcal{V} = \mathbb{R}^n$. Then $\langle x, y \rangle = x^Ty$ is the "usual" Euclidean inner product or dot product.

Example 7.12. Let $\mathcal{V} = \mathbb{R}^n$. Then $\langle x, y \rangle_Q = x^TQy$, where $Q = Q^T > 0$ is an arbitrary $n \times n$ positive definite matrix, defines a "weighted" inner product.

Definition 7.13. *If $A \in \mathbb{R}^{m \times n}$, then $A^T \in \mathbb{R}^{n \times m}$ is the unique linear transformation or map such that $\langle x, Ay \rangle = \langle A^Tx, y \rangle$ for all $x \in \mathbb{R}^m$ and for all $y \in \mathbb{R}^n$.*

It is easy to check that, with this more "abstract" definition of transpose, and if the (i, j)th element of A is a_{ij}, then the (i, j)th element of A^T is a_{ji}. It can also be checked that all the usual properties of the transpose hold, such as $(AB)^T = B^T A^T$. However, the definition above allows us to extend the concept of transpose to the case of weighted inner products in the following way. Suppose $A \in \mathbb{R}^{m \times n}$ and let $\langle \cdot, \cdot \rangle_Q$ and $\langle \cdot, \cdot \rangle_R$, with Q and R positive definite, be weighted inner products on $\mathbb{R}^m$ and $\mathbb{R}^n$, respectively. Then we can define the "weighted transpose" $A^\#$ as the unique map that satisfies

$$\langle x, Ay \rangle_Q = \langle A^\# x, y \rangle_R \quad \text{for all } x \in \mathbb{R}^m \text{ and for all } y \in \mathbb{R}^n.$$

By Example 7.12 above, we must then have $x^T QAy = x^T (A^\#)^T Ry$ for all x, y. Hence we must have $QA = (A^\#)^T R$. Taking transposes (of the usual variety) gives $A^T Q = RA^\#$. Since R is nonsingular, we find

$$A^\# = R^{-1} A^T Q.$$

We can also generalize the notion of orthogonality ($x^T y = 0$) to **Q-orthogonality** (Q is a positive definite matrix). Two vectors $x, y \in \mathbb{R}^n$ are Q-orthogonal (or conjugate with respect to Q) if $\langle x, y \rangle_Q = x^T Qy = 0$. Q-orthogonality is an important tool used in studying conjugate direction methods in optimization theory.

Definition 7.14. *Let $\mathcal{V}$ be a vector space over $\mathbb{C}$. Then $\langle \cdot, \cdot \rangle : \mathcal{V} \times \mathcal{V} \to \mathbb{C}$ is a* **complex inner product** *if*

1. $\langle x, x \rangle \geq 0$ *for all* $x \in \mathcal{V}$ *and* $\langle x, x \rangle = 0$ *if and only if* $x = 0$.
2. $\langle x, y \rangle = \overline{\langle y, x \rangle}$ *for all* $x, y \in \mathcal{V}$.
3. $\langle x, \alpha y_1 + \beta y_2 \rangle = \alpha \langle x, y_1 \rangle + \beta \langle x, y_2 \rangle$ *for all* $x, y_1, y_2 \in \mathcal{V}$ *and for all* $\alpha, \beta \in \mathbb{C}$.

Remark 7.15. We could use the notation $\langle \cdot, \cdot \rangle_c$ to denote a complex inner product, but if the vectors involved are complex-valued, the complex inner product is to be understood. Note, too, from part 2 of the definition, that $\langle x, x \rangle$ must be real for all x.

Remark 7.16. Note from parts 2 and 3 of Definition 7.14 that we have

$$\langle \alpha x_1 + \beta x_2, y \rangle = \overline{\alpha} \langle x_1, y \rangle + \overline{\beta} \langle x_2, y \rangle.$$

Remark 7.17. The Euclidean inner product of $x, y \in \mathbb{C}^n$ is given by

$$\langle x, y \rangle = \sum_{i=1}^{n} \overline{x_i} y_i = x^H y.$$

The conventional definition of the complex Euclidean inner product is $\langle x, y \rangle = y^H x$ but we use its complex conjugate $x^H y$ here for symmetry with the real case.

Remark 7.18. A weighted inner product can be defined as in the real case by $\langle x, y \rangle_Q = x^H Qy$, for arbitrary $Q = Q^H > 0$. The notion of Q-orthogonality can be similarly generalized to the complex case.

Definition 7.19. *A vector space* $(\mathcal{V}, \mathbb{F})$ *endowed with a specific inner product is called an* **inner product space**. *If* $\mathbb{F} = \mathbb{C}$, *we call* $\mathcal{V}$ *a* **complex inner product space**. *If* $\mathbb{F} = \mathbb{R}$, *we call* $\mathcal{V}$ *a* **real inner product space**.

Example 7.20.

1. Check that $\mathcal{V} = \mathbb{R}^{n\times n}$ with the inner product $\langle A, B\rangle = \operatorname{Tr} A^T B$ is a real inner product space. Note that other choices are possible since by properties of the trace function, $\operatorname{Tr} A^T B = \operatorname{Tr} B^T A = \operatorname{Tr} AB^T = \operatorname{Tr} BA^T$.

2. Check that $\mathcal{V} = \mathbb{C}^{n\times n}$ with the inner product $\langle A, B\rangle = \operatorname{Tr} A^H B$ is a complex inner product space. Again, other choices are possible.

Definition 7.21. *Let* $\mathcal{V}$ *be an inner product space. For* $v \in \mathcal{V}$, *we define the* **norm** *(or length) of* v *by* $\|v\| = \sqrt{\langle v, v\rangle}$. *This is called the norm* **induced** *by* $\langle\cdot,\cdot\rangle$.

Example 7.22.

1. If $\mathcal{V} = \mathbb{R}^n$ with the usual inner product, the induced norm is given by $\|v\| = (\sum_{i=1}^n v_i^2)^{\frac{1}{2}}$.

2. If $\mathcal{V} = \mathbb{C}^n$ with the usual inner product, the induced norm is given by $\|v\| = (\sum_{i=1}^n |v_i|^2)^{\frac{1}{2}}$.

Theorem 7.23. *Let* P *be an orthogonal projection on an inner product space* $\mathcal{V}$. *Then* $\|Pv\| \le \|v\|$ *for all* $v \in \mathcal{V}$.

Proof: Since P is an orthogonal projection, $P^2 = P = P^\#$. (Here, the notation $P^\#$ denotes the unique linear transformation that satisfies $\langle Pu, v\rangle = \langle u, P^\# v\rangle$ for all $u, v \in \mathcal{V}$. If this seems a little too abstract, consider $\mathcal{V} = \mathbb{R}^n$ (or $\mathbb{C}^n$), where $P^\#$ is simply the usual P^T (or P^H)). Hence $\langle Pv, v\rangle = \langle P^2 v, v\rangle = \langle Pv, P^\# v\rangle = \langle Pv, Pv\rangle = \|Pv\|^2 \ge 0$. Now $I - P$ is also a projection, so the above result applies and we get

$$\begin{aligned} 0 \le \langle (I-P)v, v\rangle &= \langle v, v\rangle - \langle Pv, v\rangle \\ &= \|v\|^2 - \|Pv\|^2 \end{aligned}$$

from which the theorem follows. □

Definition 7.24. *The norm induced on an inner product space by the "usual" inner product is called the* **natural norm**.

In case $\mathcal{V} = \mathbb{C}^n$ or $\mathcal{V} = \mathbb{R}^n$, the natural norm is also called the **Euclidean norm**. In the next section, other norms on these vector spaces are defined. A converse to the above procedure is also available. That is, given a norm defined by $\|x\| = \sqrt{\langle x, x\rangle}$, an inner product can be defined via the following.

Theorem 7.25 (Polarization Identity).

1. *For $x, y \in \mathbb{R}^n$, an inner product is defined by*

$$\langle x, y\rangle = x^T y = \frac{\|x+y\|^2 - \|x-y\|^2}{4} \equiv \frac{\|x+y\|^2 - \|x\|^2 - \|y\|^2}{2}.$$

2. *For $x, y \in \mathbb{C}^n$, an inner product is defined by*

$$\langle x, y\rangle = x^H y = \frac{1}{4}\sum_{k=1}^{4} j^k \|y + j^k x\|^2,$$

where $j = i = \sqrt{-1}$.

7.3 Vector Norms

Definition 7.26. *Let $(\mathcal{V}, \mathbb{F})$ be a vector space. Then $\|\cdot\| : \mathcal{V} \to \mathbb{R}$ is a* **vector norm** *if it satisfies the following three properties:*

1. *$\|x\| \geq 0$ for all $x \in \mathcal{V}$ and $\|x\| = 0$ if and only if $x = 0$.*
2. *$\|\alpha x\| = |\alpha|\|x\|$ for all $x \in \mathcal{V}$ and for all $\alpha \in \mathbb{F}$.*
3. *$\|x + y\| \leq \|x\| + \|y\|$ for all $x, y \in \mathcal{V}$.*
   (This is called the **triangle inequality***, as seen readily from the usual diagram illustrating the sum of two vectors in $\mathbb{R}^2$.)*

Remark 7.27. It is convenient in the remainder of this section to state results for complex-valued vectors. The specialization to the real case is obvious.

Definition 7.28. *A vector space $(\mathcal{V}, \mathbb{F})$ is said to be a* **normed linear space** *if and only if there exists a vector norm $\|\cdot\| : \mathcal{V} \to \mathbb{R}$ satisfying the three conditions of Definition 7.26.*

Example 7.29.

1. For $x \in \mathbb{C}^n$, the Hölder norms, or p-norms, are defined by

$$\|x\|_p = (|x_1|^p + \cdots + |x_n|^p)^{\frac{1}{p}}; \quad 1 \leq p \leq +\infty.$$

   Special cases:

   (a) $\|x\|_1 = \sum_{i=1}^n |x_i|$ (the "Manhattan" norm).

   (b) $\|x\|_2 = (\sum_{i=1}^n |x_i|^2)^{\frac{1}{2}} = (x^H x)^{\frac{1}{2}}$ (the Euclidean norm).

   (c) $\|x\|_\infty = \max_{i \in \underline{n}} |x_i| = \lim_{p \to +\infty} \|x\|_p$.
   (The second equality is a theorem that requires proof.)

2. Some weighted p-norms:

   (a) $\|x\|_{1,D} = \sum_{i=1}^{n} d_i |x_i|$, where $d_i > 0$.

   (b) $\|x\|_{2,Q} = (x^H Q x)^{\frac{1}{2}}$, where $Q = Q^H > 0$ (this norm is more commonly denoted $\|\cdot\|_Q$).

3. On the vector space $(C[t_0, t_1], \mathbb{R})$, define the vector norm

$$\|f\| = \max_{t_0 \leq t \leq t_1} |f(t)|.$$

   On the vector space $((C[t_0, t_1])^n, \mathbb{R})$, define the vector norm

$$\|f\|_\infty = \max_{t_0 \leq t \leq t_1} \|f(t)\|_\infty.$$

Theorem 7.30 (Hölder Inequality). *Let $x, y \in \mathbb{C}^n$. Then*

$$|\langle x, y\rangle| = |x^H y| \leq \|x\|_p \|y\|_q \, ; \quad \frac{1}{p} + \frac{1}{q} = 1.$$

A particular case of the Hölder inequality is of special interest.

Theorem 7.31 (Cauchy–Bunyakovsky–Schwarz Inequality). *Let $x, y \in \mathbb{C}^n$. Then*

$$|\langle x, y\rangle| = |x^H y| \leq \|x\|_2 \|y\|_2$$

with equality if and only if x and y are linearly dependent.

Proof: Consider the matrix $[x \;\; y] \in \mathbb{C}^{n \times 2}$. Since

$$[x \;\; y]^H [x \;\; y] = \begin{bmatrix} x^H x & x^H y \\ y^H x & y^H y \end{bmatrix}$$

is a nonnegative definite matrix, its determinant must be nonnegative. In other words, $0 \leq (x^H x)(y^H y) - (x^H y)(y^H x)$. Since $y^H x = \overline{x^H y}$, we see immediately that $|x^H y| \leq \|x\|_2 \|y\|_2$. □

Note: This is not the classical algebraic proof of the Cauchy–Bunyakovsky–Schwarz (C-B-S) inequality (see, e.g., [20, p. 217]). However, it is particularly easy to remember.

Remark 7.32. The **angle** θ between two nonzero vectors $x, y \in \mathbb{C}^n$ may be defined by $\cos\theta = \frac{|x^H y|}{\|x\|_2 \|y\|_2}$, $0 \leq \theta \leq \frac{\pi}{2}$. The C-B-S inequality is thus equivalent to the statement $|\cos\theta| \leq 1$.

Remark 7.33. Theorem 7.31 and Remark 7.32 are true for general inner product spaces.

Remark 7.34. The norm $\|\cdot\|_2$ is **unitarily invariant**, i.e., if $U \in \mathbb{C}^{n \times n}$ is unitary, then $\|Ux\|_2 = \|x\|_2$ (*Proof*: $\|Ux\|_2^2 = x^H U^H U x = x^H x = \|x\|_2^2$). However, $\|\cdot\|_1$ and $\|\cdot\|_\infty$

are not unitarily invariant. Similar remarks apply to the unitary invariance of norms of real vectors under orthogonal transformation.

Remark 7.35. If $x, y \in \mathbb{C}^n$ are orthogonal, then we have the **Pythagorean Identity**

$$\|x \pm y\|_2^2 = \|x\|_2^2 + \|y\|_2^2,$$

the proof of which follows easily from $\|z\|_2^2 = z^H z$.

Theorem 7.36. *All norms on $\mathbb{C}^n$ are* **equivalent***; i.e., there exist constants c_1, c_2 (possibly depending on n) such that*

$$c_1\|x\|_\alpha \leq \|x\|_\beta \leq c_2\|x\|_\alpha \quad \textit{for all } x \in \mathbb{C}^n.$$

Example 7.37. For $x \in \mathbb{C}^n$, the following inequalities are all tight bounds; i.e., there exist vectors x for which equality holds:

$$\begin{array}{ll} \|x\|_1 \leq \sqrt{n}\,\|x\|_2, & \|x\|_1 \leq n\,\|x\|_\infty; \\ \|x\|_2 \leq \|x\|_1, & \|x\|_2 \leq \sqrt{n}\,\|x\|_\infty; \\ \|x\|_\infty \leq \|x\|_1, & \|x\|_\infty \leq \|x\|_2. \end{array}$$

Finally, we conclude this section with a theorem about convergence of vectors. Convergence of a sequence of vectors to some limit vector can be converted into a statement about convergence of real numbers, i.e., convergence in terms of vector norms.

Theorem 7.38. *Let $\|\cdot\|$ be a vector norm and suppose $v, v^{(1)}, v^{(2)}, \ldots \in \mathbb{C}^n$. Then*

$$\lim_{k\to+\infty} v^{(k)} = v \quad \textit{if and only if} \quad \lim_{k\to+\infty} \|v^{(k)} - v\| = 0.$$

7.4 Matrix Norms

In this section we introduce the concept of matrix norm. As with vectors, the motivation for using matrix norms is to have a notion of either the size of or the nearness of matrices. The former notion is useful for perturbation analysis, while the latter is needed to make sense of "convergence" of matrices. Attention is confined to the vector space $(\mathbb{R}^{m\times n}, \mathbb{R})$ since that is what arises in the majority of applications. Extension to the complex case is straightforward and essentially obvious.

Definition 7.39. $\|\cdot\| : \mathbb{R}^{m\times n} \to \mathbb{R}$ *is a* **matrix norm** *if it satisfies the following three properties:*

1. $\|A\| \geq 0$ *for all* $A \in \mathbb{R}^{m\times n}$ *and* $\|A\| = 0$ *if and only if* $A = 0$.
2. $\|\alpha A\| = |\alpha|\|A\|$ *for all* $A \in \mathbb{R}^{m\times n}$ *and for all* $\alpha \in \mathbb{R}$.
3. $\|A + B\| \leq \|A\| + \|B\|$ *for all* $A, B \in \mathbb{R}^{m\times n}$.
   (As with vectors, this is called the **triangle inequality***.)*

Example 7.40. Let $A \in \mathbb{R}^{m \times n}$. Then the **Frobenius norm** (or matrix Euclidean norm) is defined by

$$\|A\|_F = \left(\sum_{i=1}^{m} \sum_{j=1}^{n} a_{ij}^2 \right)^{\frac{1}{2}} = \left(\sum_{k=1}^{r} \sigma_k^2(A) \right)^{\frac{1}{2}} = (\mathrm{Tr}\,(A^T A))^{\frac{1}{2}} = (\mathrm{Tr}\,(AA^T))^{\frac{1}{2}}$$

(where $r = \operatorname{rank}(A)$).

Example 7.41. Let $A \in \mathbb{R}^{m \times n}$. Then the **matrix** ***p*****-norms** are defined by

$$\|A\|_p = \max_{\|x\|_p \neq 0} \frac{\|Ax\|_p}{\|x\|_p} = \max_{\|x\|_p = 1} \|Ax\|_p \,.$$

The following three special cases are important because they are "computable." Each is a theorem and requires a proof.

1. The "maximum column sum" norm is

$$\|A\|_1 = \max_{j \in \underline{n}} \left(\sum_{i=1}^{m} |a_{ij}| \right).$$

2. The "maximum row sum" norm is

$$\|A\|_\infty = \max_{i \in \underline{m}} \left(\sum_{j=1}^{n} |a_{ij}| \right).$$

3. The spectral norm is

$$\|A\|_2 = \lambda_{\max}^{\frac{1}{2}}(A^T A) = \lambda_{\max}^{\frac{1}{2}}(AA^T) = \sigma_1(A).$$

   Note: $\|A^+\|_2 = 1/\sigma_r(A)$, where $r = \operatorname{rank}(A)$.

Example 7.42. Let $A \in \mathbb{R}^{m \times n}$. The **Schatten** ***p*****-norms** are defined by

$$\|A\|_{S,p} = (\sigma_1^p + \cdots + \sigma_r^p)^{\frac{1}{p}}.$$

Some special cases of Schatten p-norms are equal to norms defined previously. For example, $\|\cdot\|_{S,2} = \|\cdot\|_F$ and $\|\cdot\|_{S,\infty} = \|\cdot\|_2$. The norm $\|\cdot\|_{S,1}$ is often called the trace norm.

Example 7.43. Let $A \in \mathbb{R}^{m \times n}$. Then "mixed" norms can also be defined by

$$\|A\|_{p,q} = \max_{\|x\|_q \neq 0} \frac{\|Ax\|_p}{\|x\|_q}.$$

Example 7.44. The "matrix analogue of the vector 1-norm," $\|A\|_s = \sum_{i,j} |a_{ij}|$, is a norm.

The concept of a matrix norm alone is not altogether useful since it does not allow us to estimate the size of a matrix product AB in terms of the sizes of A and B individually.

Notice that this difficulty did not arise for vectors, although there are analogues for, e.g., inner products or outer products of vectors. We thus need the following definition.

Definition 7.45. *Let $A \in \mathbb{R}^{m \times n}$, $B \in \mathbb{R}^{n \times k}$. Then the norms $\|\cdot\|_\alpha$, $\|\cdot\|_\beta$, and $\|\cdot\|_\gamma$ are* **mutually consistent** *if $\|AB\|_\alpha \leq \|A\|_\beta \|B\|_\gamma$. A matrix norm $\|\cdot\|$ is said to be* **consistent** *if $\|AB\| \leq \|A\|\|B\|$ whenever the matrix product is defined.*

Example 7.46.

1. $\|\cdot\|_F$ and $\|\cdot\|_p$ for all p are consistent matrix norms.

2. The "mixed" norm

$$\|\cdot\|_{1,\infty} = \max_{x \neq 0} \frac{\|Ax\|_1}{\|x\|_\infty} = \max_{i,j} |a_{ij}|$$

is a matrix norm but it is not consistent. For example, take $A = B = \left[\begin{smallmatrix} 1 & 1 \\ 1 & 1 \end{smallmatrix}\right]$. Then $\|AB\|_{1,\infty} = 2$ while $\|A\|_{1,\infty}\|B\|_{1,\infty} = 1$.

The p-norms are examples of matrix norms that are **subordinate to (or induced by) a vector norm**, i.e.,

$$\|A\| = \max_{x \neq 0} \frac{\|Ax\|}{\|x\|} = \max_{\|x\|=1} \|Ax\|$$

(or, more generally, $\|A\|_{p,q} = \max_{x \neq 0} \frac{\|Ax\|_p}{\|x\|_q}$). For such subordinate norms, also called **operator norms**, we clearly have $\|Ax\| \leq \|A\|\|x\|$. Since $\|ABx\| \leq \|A\|\|Bx\| \leq \|A\|\|B\|\|x\|$, it follows that all subordinate norms are consistent.

Theorem 7.47. *There exists a vector x^* such that $\|Ax^*\| = \|A\|\|x^*\|$ if the matrix norm is subordinate to the vector norm.*

Theorem 7.48. *If $\|\cdot\|_m$ is a consistent matrix norm, there exists a vector norm $\|\cdot\|_v$ consistent with it, i.e., $\|Ax\|_v \leq \|A\|_m \|x\|_v$.*

Not every consistent matrix norm is subordinate to a vector norm. For example, consider $\|\cdot\|_F$. Then $\|Ax\|_2 \leq \|A\|_F\|x\|_2$, so $\|\cdot\|_2$ is consistent with $\|\cdot\|_F$, but there does **not** exist a vector norm $\|\cdot\|$ such that $\|A\|_F$ is given by $\max_{x \neq 0} \frac{\|Ax\|}{\|x\|}$.

Useful Results

The following miscellaneous results about matrix norms are collected for future reference. The interested reader is invited to prove each of them as an exercise.

1. $\|I_n\|_p = 1$ for all p, while $\|I_n\|_F = \sqrt{n}$.

2. For $A \in \mathbb{R}^{n \times n}$, the following inequalities are all tight, i.e., there exist matrices A for which equality holds:

$$\begin{array}{lll} \|A\|_1 \leq \sqrt{n}\,\|A\|_2, & \|A\|_1 \leq n\,\|A\|_\infty, & \|A\|_1 \leq \sqrt{n}\,\|A\|_F; \\ \|A\|_2 \leq \sqrt{n}\,\|A\|_1, & \|A\|_2 \leq \sqrt{n}\,\|A\|_\infty, & \|A\|_2 \leq \|A\|_F; \\ \|A\|_\infty \leq n\,\|A\|_1, & \|A\|_\infty \leq \sqrt{n}\,\|A\|_2, & \|A\|_\infty \leq \sqrt{n}\,\|A\|_F; \\ \|A\|_F \leq \sqrt{n}\,\|A\|_1, & \|A\|_F \leq \sqrt{n}\,\|A\|_2, & \|A\|_F \leq \sqrt{n}\,\|A\|_\infty. \end{array}$$

3. For $A \in \mathbb{R}^{m \times n}$,

$$\max_{i,j} |a_{ij}| \leq \|A\|_2 \leq \sqrt{mn} \max_{i,j} |a_{ij}|.$$

4. The norms $\|\cdot\|_F$ and $\|\cdot\|_2$ (as well as all the Schatten p-norms, but not necessarily other p-norms) are **unitarily invariant**; i.e., for all $A \in \mathbb{R}^{m \times n}$ and for all orthogonal matrices $Q \in \mathbb{R}^{m \times m}$ and $Z \in \mathbb{R}^{n \times n}$, $\|QAZ\|_\alpha = \|A\|_\alpha$ for $\alpha = 2$ or F.

Convergence

The following theorem uses matrix norms to convert a statement about convergence of a sequence of matrices into a statement about the convergence of an associated sequence of scalars.

Theorem 7.49. *Let $\|\cdot\|$ be a matrix norm and suppose $A, A^{(1)}, A^{(2)}, \ldots \in \mathbb{R}^{m \times n}$. Then*

$$\lim_{k \to +\infty} A^{(k)} = A \text{ if and only if } \lim_{k \to +\infty} \|A^{(k)} - A\| = 0.$$

EXERCISES

1. If P is an orthogonal projection, prove that $P^+ = P$.

2. Suppose P and Q are orthogonal projections and $P + Q = I$. Prove that $P - Q$ must be an orthogonal matrix.

3. Prove that $I - A^+ A$ is an orthogonal projection. Also, prove directly that $V_2 V_2^T$ is an orthogonal projection, where V_2 is defined as in Theorem 5.1.

4. Suppose that a matrix $A \in \mathbb{R}^{m \times n}$ has linearly independent columns. Prove that the orthogonal projection onto the space spanned by these column vectors is given by the matrix $P = A(A^T A)^{-1} A^T$.

5. Find the (orthogonal) projection of the vector $[2 \ \ 3 \ \ 4]^T$ onto the subspace of $\mathbb{R}^3$ spanned by the plane $3x - y + 2z = 0$.

6. Prove that $\mathbb{R}^{n \times n}$ with the inner product $\langle A, B \rangle = \operatorname{Tr} A^T B$ is a real inner product space.

7. Show that the matrix norms $\|\cdot\|_2$ and $\|\cdot\|_F$ are unitarily invariant.

8. **Definition:** Let $A \in \mathbb{R}^{n \times n}$ and denote its set of eigenvalues (not necessarily distinct) by $\{\lambda_1, \ldots, \lambda_n\}$. The *spectral radius* of A is the scalar

$$\rho(A) = \max_i |\lambda_i|.$$

Let

$$A = \begin{bmatrix} 0 & 1 & 0 \\ 0 & 0 & 1 \\ 14 & 12 & 5 \end{bmatrix}.$$

Determine $\|A\|_F$, $\|A\|_1$, $\|A\|_2$, $\|A\|_\infty$, and $\rho(A)$.

9. Let

$$A = \begin{bmatrix} 8 & 1 & 6 \\ 3 & 5 & 7 \\ 4 & 9 & 2 \end{bmatrix}.$$

Determine $\|A\|_F$, $\|A\|_1$, $\|A\|_2$, $\|A\|_\infty$, and $\rho(A)$. (An $n \times n$ matrix, all of whose columns and rows as well as main diagonal and antidiagonal sum to $s = n(n^2+1)/2$, is called a "magic square" matrix. If M is a magic square matrix, it can be proved that $\|M\|_p = s$ for all p.)

10. Let $A = xy^T$, where both $x, y \in \mathbb{R}^n$ are nonzero. Determine $\|A\|_F$, $\|A\|_1$, $\|A\|_2$, and $\|A\|_\infty$ in terms of $\|x\|_\alpha$ and/or $\|y\|_\beta$, where α and β take the value 1, 2, or ∞ as appropriate.

Chapter 8

Linear Least Squares Problems

8.1 The Linear Least Squares Problem

Problem: Suppose $A \in \mathbb{R}^{m \times n}$ with $m \geq n$ and $b \in \mathbb{R}^m$ is a given vector. The **linear least squares problem** consists of finding an element of the set

$$\mathcal{X} = \{x \in \mathbb{R}^n : \rho(x) = \|Ax - b\|_2 \text{ is minimized}\}.$$

Solution: The set $\mathcal{X}$ has a number of easily verified properties:

1. A vector $x \in \mathcal{X}$ if and only if $A^T r = 0$, where $r = b - Ax$ is the **residual** associated with x. The equations $A^T r = 0$ can be rewritten in the form $A^T A x = A^T b$ and the latter form is commonly known as the **normal equations**, i.e., $x \in \mathcal{X}$ if and only if x is a solution of the normal equations. For further details, see Section 8.2.

2. A vector $x \in \mathcal{X}$ if and only if x is of the form

$$x = A^+ b + (I - A^+ A)y, \quad \text{where } y \in \mathbb{R}^n \text{ is arbitrary.} \tag{8.1}$$

To see why this must be so, write the residual r in the form

$$r = (b - P_{\mathcal{R}(A)} b) + (P_{\mathcal{R}(A)} b - Ax).$$

Now, $(P_{\mathcal{R}(A)} b - Ax)$ is clearly in $\mathcal{R}(A)$, while

$$\begin{aligned}(b - P_{\mathcal{R}(A)} b) &= (I - P_{\mathcal{R}(A)}) b \\ &= P_{\mathcal{R}(A)^\perp} b \in \mathcal{R}(A)^\perp\end{aligned}$$

so these two vectors are orthogonal. Hence,

$$\begin{aligned}\|r\|_2^2 &= \|b - Ax\|_2^2 \\ &= \|b - P_{\mathcal{R}(A)} b\|_2^2 + \|P_{\mathcal{R}(A)} b - Ax\|_2^2\end{aligned}$$

from the Pythagorean identity (Remark 7.35). Thus, $\|Ax - b\|_2^2$ (and hence $\rho(x) = \|Ax - b\|_2$) assumes its minimum value if and only if

$$Ax = P_{\mathcal{R}(A)} b = AA^+ b \tag{8.2}$$

and this equation always has a solution since $AA^+b \in \mathcal{R}(A)$. By Theorem 6.3, all solutions of (8.2) are of the form

$$\begin{aligned} x &= A^+AA^+b + (I - A^+A)y \\ &= A^+b + (I - A^+A)y, \end{aligned}$$

where $y \in \mathbb{R}^n$ is arbitrary. The minimum value of $\rho(x)$ is then clearly equal to

$$\begin{aligned} \|b - P_{\mathcal{R}(A)}b\|_2 &= \|(I - AA^+)b\|_2 \\ &\leq \|b\|_2, \end{aligned}$$

the last inequality following by Theorem 7.23.

3. $\mathcal{X}$ is convex. To see why, consider two arbitrary vectors $x_1 = A^+b + (I - A^+A)y$ and $x_2 = A^+b + (I - A^+A)z$ in $\mathcal{X}$. Let $\theta \in [0, 1]$. Then the convex combination $\theta x_1 + (1 - \theta)x_2 = A^+b + (I - A^+A)(\theta y + (1 - \theta)z)$ is clearly in $\mathcal{X}$.

4. $\mathcal{X}$ has a unique element x^* of minimal 2-norm. In fact, $x^* = A^+b$ is the unique vector that solves this "double minimization" problem, i.e., x^* minimizes the residual $\rho(x)$ and is the vector of minimum 2-norm that does so. This follows immediately from convexity or directly from the fact that all $x \in \mathcal{X}$ are of the form (8.1) and

$$\|x\|_2^2 = \|x^*\|_2^2 + \|(I - A^+A)y\|_2^2,$$

which follows since the two vectors are orthogonal.

5. There is a unique solution to the least squares problem, i.e., $\mathcal{X} = \{x^*\} = \{A^+b\}$, if and only if $A^+A = I$ or, equivalently, if and only if $\text{rank}(A) = n$.

Just as for the solution of linear equations, we can generalize the linear least squares problem to the matrix case.

Theorem 8.1. *Let $A \in \mathbb{R}^{m \times n}$ and $B \in \mathbb{R}^{m \times k}$. The general solution to*

$$\min_{X \in \mathbb{R}^{n \times k}} \|AX - B\|_2$$

is of the form

$$X = A^+B + (I - A^+A)Y,$$

where $Y \in \mathbb{R}^{n \times k}$ is arbitrary. The unique solution of minimum 2-norm or F-norm is $X = A^+B$.

Remark 8.2. Notice that solutions of the linear least squares problem look exactly the same as solutions of the linear system $AX = B$. The only difference is that in the case of linear least squares solutions, there is no "existence condition" such as $\mathcal{R}(B) \subseteq \mathcal{R}(A)$. If the existence condition happens to be satisfied, then equality holds and the least squares residual is 0. Of all solutions that give a residual of 0, the unique solution $X = A^+B$ has minimum 2-norm or F-norm.

Remark 8.3. If we take $B = I_m$ in Theorem 8.1, then $X = A^+$ can be interpreted as saying that the Moore–Penrose pseudoinverse of A is the best (in the matrix 2-norm sense) matrix such that AX approximates the identity.

Remark 8.4. Many other interesting and useful approximation results are available for the matrix 2-norm (and F-norm). One such is the following. Let $A \in \mathbb{R}_r^{m\times n}$ with SVD

$$A = U\Sigma V^T = \sum_{i=1}^{r} \sigma_i u_i v_i^T.$$

Then a best rank k approximation to A for $1 \leq k \leq r$, i.e., a solution to

$$\min_{M\in\mathbb{R}_k^{m\times n}} \|A - M\|_2,$$

is given by

$$M_k = \sum_{i=1}^{k} \sigma_i u_i v_i^T.$$

The special case in which $m = n$ and $k = n-1$ gives a nearest singular matrix to $A \in \mathbb{R}_n^{n\times n}$.

8.2 Geometric Solution

Looking at the schematic provided in Figure 8.1, it is apparent that minimizing $\|Ax - b\|_2$ is equivalent to finding the vector $x \in \mathbb{R}^n$ for which $p = Ax$ is closest to b (in the Euclidean norm sense). Clearly, $r = b - Ax$ must be orthogonal to $\mathcal{R}(A)$. Thus, if Ay is an arbitrary vector in $\mathcal{R}(A)$ (i.e., y is arbitrary), we must have

$$\begin{aligned} 0 &= (Ay)^T(b - Ax) \\ &= y^TA^T(b - Ax) \\ &= y^T(A^Tb - A^TAx). \end{aligned}$$

Since y is arbitrary, we must have $A^Tb - A^TAx = 0$ or $A^TAx = A^Tb$.
Special case: If A is full (column) rank, then $x = (A^TA)^{-1}A^Tb$.

8.3 Linear Regression and Other Linear Least Squares Problems

8.3.1 Example: Linear regression

Suppose we have m measurements $(t_1, y_1), \ldots, (t_m, y_m)$ for which we hypothesize a linear (affine) relationship

$$y = \alpha t + \beta \tag{8.3}$$

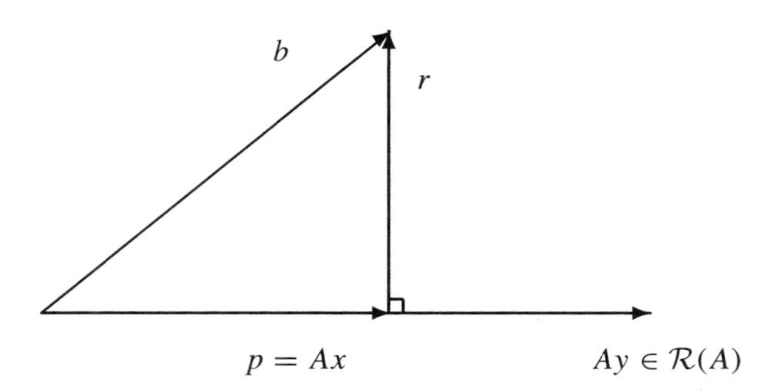

Figure 8.1. *Projection of* b *on* $\mathcal{R}(A)$.

for certain constants α and β. One way to solve this problem is to find the line that best fits the data in the least squares sense; i.e., with the model (8.3), we have

$$\begin{aligned} y_1 &= \alpha t_1 + \beta + \delta_1, \\ y_2 &= \alpha t_2 + \beta + \delta_2 \\ \vdots &\quad \vdots \\ y_m &= \alpha t_m + \beta + \delta_m, \end{aligned}$$

where $\delta_1, \ldots, \delta_m$ are "errors" and we wish to minimize $\delta_1^2 + \cdots + \delta_m^2$. Geometrically, we are trying to find the best line that minimizes the (sum of squares of the) distances from the given data points. See, for example, Figure 8.2.

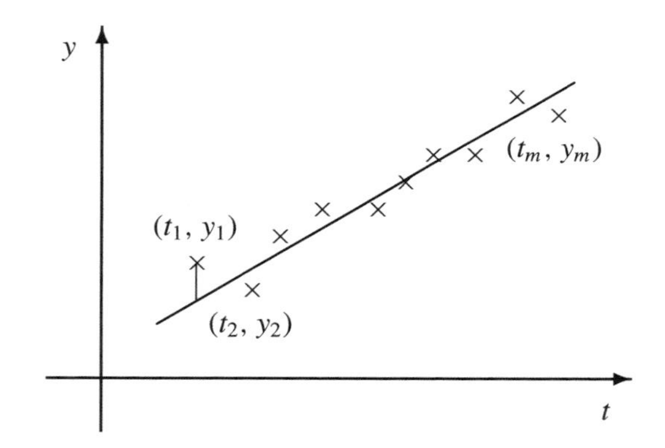

Figure 8.2. *Simple linear regression.*

Note that distances are measured in the vertical sense from the points to the line (as indicated, for example, for the point (t_1, y_1)). However, other criteria are possible. For example, one could measure the distances in the horizontal sense, or the perpendicular distance from the points to the line could be used. The latter is called **total least squares**. Instead of 2-norms, one could also use 1-norms or ∞-norms. The latter two are computationally much more difficult to handle, and thus we present only the more tractable 2-norm case in text that follows.

The m "error equations" can be written in matrix form as

$$y = Ax + \delta,$$

where

$$y = \begin{bmatrix} y_1 \\ \vdots \\ y_m \end{bmatrix}, \quad \delta = \begin{bmatrix} \delta_1 \\ \vdots \\ \delta_m \end{bmatrix}, \quad A = \begin{bmatrix} t_1 & 1 \\ \vdots & \vdots \\ t_m & 1 \end{bmatrix}, \text{ and } x = \begin{bmatrix} \alpha \\ \beta \end{bmatrix}.$$

We then want to solve the problem

$$\min \delta^T \delta = \min_x (Ax - y)^T (Ax - y)$$

or, equivalently,

$$\min \|\delta\|_2^2 = \min_x \|Ax - y\|_2^2. \tag{8.4}$$

Solution: $x = \left[\begin{smallmatrix} \alpha \\ \beta \end{smallmatrix}\right]$ is a solution of the normal equations $A^T A x = A^T y$ where, for the special form of the matrices above, we have

$$A^T A = \begin{bmatrix} \sum_i t_i^2 & \sum_i t_i \\ \sum_i t_i & m \end{bmatrix}$$

and

$$A^T y = \begin{bmatrix} \sum_i t_i y_i \\ \sum_i y_i \end{bmatrix}.$$

The solution for the parameters α and β can then be written

$$x = (A^T A)^{-1} A^T y = \frac{1}{m \sum_i t_i^2 - (\sum_i t_i)^2} \begin{bmatrix} m \sum_i t_i y_i - \sum_i t_i \sum_i y_i \\ \sum_i t_i^2 \sum_i y_i - \sum_i t_i \sum_i t_i y_i \end{bmatrix}.$$

8.3.2 Other least squares problems

Suppose the hypothesized model is not the linear equation (8.3) but rather is of the form

$$y = f(t) = c_1 \phi_1(t) + \cdots + c_n \phi_n(t). \tag{8.5}$$

In (8.5) the $\phi_i(t)$ are given (basis) functions and the c_i are constants to be determined to minimize the least squares error. The matrix problem is still (8.4), where we now have

$$A = \begin{bmatrix} \phi_1(t_1) & \cdots & \phi_n(t_1) \\ \vdots & & \vdots \\ \phi_1(t_m) & \cdots & \phi_n(t_m) \end{bmatrix}, \quad y = \begin{bmatrix} y_1 \\ \vdots \\ y_m \end{bmatrix}, \text{ and } x = \begin{bmatrix} c_1 \\ \vdots \\ c_n \end{bmatrix}.$$

An important special case of (8.5) is least squares polynomial approximation, which corresponds to choosing $\phi_i(t) = t^{i-1}$, $i \in \underline{n}$, although this choice can lead to computational difficulties because of numerical ill conditioning for large n. Numerically better approaches are based on orthogonal polynomials, piecewise polynomial functions, splines, etc.

The key feature in (8.5) is that the coefficients c_i appear linearly. The basis functions ϕ_i can be arbitrarily nonlinear. Sometimes a problem in which the c_i's appear nonlinearly can be converted into a linear problem. For example, if the fitting function is of the form $y = f(t) = c_1 e^{c_2 t}$, then taking logarithms yields the equation $\log y = \log c_1 + c_2 t$. Then defining $\tilde{y} = \log y$, $\tilde{c}_1 = \log c_1$, and $\tilde{c}_2 = c_2$ results in a standard linear least squares problem.

8.4 Least Squares and Singular Value Decomposition

In the numerical linear algebra literature (e.g., [4], [7], [11], [23]), it is shown that solution of linear least squares problems via the normal equations can be a very poor numerical method in finite-precision arithmetic. Since the standard Kalman filter essentially amounts to sequential updating of normal equations, it can be expected to exhibit such poor numerical behavior in practice (and it does). Better numerical methods are based on algorithms that work directly and solely on A itself rather than $A^T A$. Two basic classes of algorithms are based on SVD and QR (orthogonal-upper triangular) factorization, respectively. The former is much more expensive but is generally more reliable and offers considerable theoretical insight.

In this section we investigate solution of the linear least squares problem

$$\min_x \|Ax - b\|_2, \quad A \in \mathbb{R}^{m\times n}, \quad b \in \mathbb{R}^m, \tag{8.6}$$

via the SVD. Specifically, we assume that A has an SVD given by $A = U\Sigma V^T = U_1 S V_1^T$ as in Theorem 5.1. We now note that

$$\begin{aligned}
\|Ax - b\|_2^2 &= \|U\Sigma V^T x - b\|_2^2 \\
&= \|\Sigma V^T x - U^T b\|_2^2 \quad \text{since } \|\cdot\|_2 \text{ is unitarily invariant} \\
&= \|\Sigma z - c\|_2^2 \quad \text{where } z = V^T x,\ c = U^T b \\
&= \left\| \begin{bmatrix} S & 0 \\ 0 & 0 \end{bmatrix} \begin{bmatrix} z_1 \\ z_2 \end{bmatrix} - \begin{bmatrix} c_1 \\ c_2 \end{bmatrix} \right\|_2^2 \\
&= \left\| \begin{bmatrix} S z_1 - c_1 \\ -c_2 \end{bmatrix} \right\|_2^2 \\
&= \|S z_1 - c_1\|_2^2 + \|c_2\|_2^2 .
\end{aligned}$$

The last equality follows from the fact that if $v = \left[\begin{smallmatrix} v_1 \\ v_2 \end{smallmatrix}\right]$, then $\|v\|_2^2 = \|v_1\|_2^2 + \|v_2\|_2^2$ (note that orthogonality is not what is used here; the subvectors can have different lengths). This explains why it is convenient to work above with the square of the norm rather than the norm. As far as the minimization is concerned, the two are equivalent. In fact, the last quantity above is clearly minimized by taking $z_1 = S^{-1} c_1$. The subvector z_2 is arbitrary, while the minimum value of $\|Ax - b\|_2^2$ is $\|c_2\|_2^2$.

Now transform back to the original coordinates:

$$
\begin{aligned}
x &= Vz \\
&= [V_1 \quad V_2] \begin{bmatrix} z_1 \\ z_2 \end{bmatrix} \\
&= V_1 z_1 + V_2 z_2 \\
&= V_1 S^{-1} c_1 + V_2 z_2 \\
&= V_1 S^{-1} U_1^T b + V_2 z_2.
\end{aligned}
$$

The last equality follows from

$$
c = U^T b = \begin{bmatrix} U_1^T b \\ U_2^T b \end{bmatrix} = \begin{bmatrix} c_1 \\ c_2 \end{bmatrix}.
$$

Note that since z_2 is arbitrary, $V_2 z_2$ is an arbitrary vector in $\mathcal{R}(V_2) = \mathcal{N}(A)$. Thus, x has been written in the form $x = A^+ b + (I - A^+ A)y$, where $y \in \mathbb{R}^m$ is arbitrary. This agrees, of course, with (8.1).

The minimum value of the least squares residual is

$$
\|c_2\|_2 = \|U_2^T b\|_2
$$

and we clearly have that

$$
\begin{aligned}
\text{minimum least squares residual is } 0 &\iff b \text{ is orthogonal to all vectors in } U_2 \\
&\iff b \text{ is orthogonal to all vectors in } \mathcal{R}(A)^\perp \\
&\iff b \in \mathcal{R}(A).
\end{aligned}
$$

Another expression for the minimum residual is $\|(I - AA^+)b\|_2$. This follows easily since $\|(I - AA^+)b\|_2^2 = \|U_2 U_2^T b\|_2^2 = b^T U_2 U_2^T U_2 U_2^T b = b^T U_2 U_2^T b = \|U_2^T b\|_2^2$.

Finally, an important special case of the linear least squares problem is the so-called full-rank problem, i.e., $A \in \mathbb{R}_n^{m \times n}$. In this case the SVD of A is given by $A = U \Sigma V^T = [U_1 \quad U_2] \left[\begin{smallmatrix} S \\ 0 \end{smallmatrix}\right] V_1^T$, and there is thus "no V_2 part" to the solution.

8.5 Least Squares and QR Factorization

In this section, we again look at the solution of the linear least squares problem (8.6) but this time in terms of the QR factorization. This matrix factorization is much cheaper to compute than an SVD and, with appropriate numerical enhancements, can be quite reliable.

To simplify the exposition, we add the simplifying assumption that A has full column rank, i.e., $A \in \mathbb{R}_n^{m \times n}$. It is then possible, via a sequence of so-called Householder or Givens transformations, to reduce A in the following way. A finite sequence of simple orthogonal row transformations (of Householder or Givens type) can be performed on A to reduce it to triangular form. If we label the product of such orthogonal row transformations as the orthogonal matrix $Q^T \in \mathbb{R}^{m \times m}$, we have

$$
Q^T A = \begin{bmatrix} R \\ 0 \end{bmatrix}, \tag{8.7}
$$

where $R \in \mathbb{R}_n^{n \times n}$ is upper triangular. Now write $Q = [Q_1 \quad Q_2]$, where $Q_1 \in \mathbb{R}^{m \times n}$ and $Q_2 \in \mathbb{R}^{m \times (m-n)}$. Both Q_1 and Q_2 have orthonormal columns. Multiplying through by Q in (8.7), we see that

$$\begin{aligned} A &= Q \begin{bmatrix} R \\ 0 \end{bmatrix} \qquad (8.8) \\ &= [Q_1 \quad Q_2] \begin{bmatrix} R \\ 0 \end{bmatrix} \\ &= Q_1 R. \qquad (8.9) \end{aligned}$$

Any of (8.7), (8.8), or (8.9) are variously referred to as QR factorizations of A. Note that (8.9) is essentially what is accomplished by the Gram–Schmidt process, i.e., by writing $AR^{-1} = Q_1$ we see that a "triangular" linear combination (given by the coefficients of R^{-1}) of the columns of A yields the orthonormal columns of Q_1.

Now note that

$$\begin{aligned} \|Ax - b\|_2^2 &= \|Q^T Ax - Q^T b\|_2^2 \quad \text{since } \|\cdot\|_2 \text{ is unitarily invariant} \\ &= \left\| \begin{bmatrix} R \\ 0 \end{bmatrix} x - \begin{bmatrix} c_1 \\ c_2 \end{bmatrix} \right\|_2^2, \quad \text{where } \begin{bmatrix} c_1 \\ c_2 \end{bmatrix} = \begin{bmatrix} Q_1^T b \\ Q_2^T b \end{bmatrix} \\ &= \|Rx - c_1\|_2^2 + \|c_2\|_2^2. \end{aligned}$$

The last quantity above is clearly minimized by taking $x = R^{-1} c_1$ and the minimum residual is $\|c_2\|_2$. Equivalently, we have $x = R^{-1} Q_1^T b = A^+ b$ and the minimum residual is $\|Q_2^T b\|_2$.

EXERCISES

1. For $A \in \mathbb{R}^{m \times n}$, $b \in \mathbb{R}^m$, and any $y \in \mathbb{R}^n$, check directly that $(I - A^+ A)y$ and $A^+ b$ are orthogonal vectors.

2. Consider the following set of measurements (x_i, y_i):

$$(1, 2), \quad (2, 1), \quad (3, 3).$$

   (a) Find the best (in the 2-norm sense) line of the form $y = \alpha x + \beta$ that fits this data.

   (b) Find the best (in the 2-norm sense) line of the form $x = \alpha y + \beta$ that fits this data.

3. Suppose q_1 and q_2 are two orthonormal vectors and b is a fixed vector, all in $\mathbb{R}^n$.

   (a) Find the optimal linear combination $\alpha q_1 + \beta q_2$ that is closest to b (in the 2-norm sense).

   (b) Let r denote the "error vector" $b - \alpha q_1 - \beta q_2$. Show that r is orthogonal to both q_1 and q_2.

4. Find all solutions of the linear least squares problem

$$\min_x \|Ax - b\|_2$$

when $A = \begin{bmatrix} 1 & 1 \\ 1 & 1 \end{bmatrix}$ and $b = \begin{bmatrix} 1 \\ 2 \end{bmatrix}$.

5. Consider the problem of finding the minimum 2-norm solution of the linear least squares problem

$$\min_x \|Ax - b\|_2$$

when $A = \begin{bmatrix} 1 & 0 \\ 0 & 0 \end{bmatrix}$ and $b = \begin{bmatrix} 1 \\ 1 \end{bmatrix}$. The solution is

$$x^* = A^+ b = \begin{bmatrix} 1 & 0 \\ 0 & 0 \end{bmatrix} \begin{bmatrix} 1 \\ 1 \end{bmatrix} = \begin{bmatrix} 1 \\ 0 \end{bmatrix}.$$

(a) Consider a perturbation $E_1 = \begin{bmatrix} 0 & \delta \\ 0 & 0 \end{bmatrix}$ of A, where δ is a small positive number. Solve the perturbed version of the above problem,

$$\min_y \|A_1 y - b\|_2,$$

where $A_1 = A + E_1$. What happens to $\|x^* - y\|_2$ as δ approaches 0?

(b) Now consider the perturbation $E_2 = \begin{bmatrix} 0 & 0 \\ 0 & \delta \end{bmatrix}$ of A, where again δ is a small positive number. Solve the perturbed problem

$$\min_z \|A_2 z - b\|_2,$$

where $A_2 = A + E_2$. What happens to $\|x^* - z\|_2$ as δ approaches 0?

6. Use the four Penrose conditions and the fact that Q_1 has orthonormal columns to verify that if $A \in \mathbb{R}_n^{m \times n}$ can be factored in the form (8.9), then $A^+ = R^{-1} Q_1^T$.

7. Let $A \in \mathbb{R}^{n \times n}$, not necessarily nonsingular, and suppose $A = QR$, where Q is orthogonal. Prove that $A^+ = R^+ Q^T$.

Chapter 9

Eigenvalues and Eigenvectors

9.1 Fundamental Definitions and Properties

Definition 9.1. *A nonzero vector $x \in \mathbb{C}^n$ is a* **right eigenvector** *of $A \in \mathbb{C}^{n\times n}$ if there exists a scalar $\lambda \in \mathbb{C}$, called an* **eigenvalue**, *such that*

$$Ax = \lambda x. \tag{9.1}$$

Similarly, a nonzero vector $y \in \mathbb{C}^n$ is a **left eigenvector** *corresponding to an eigenvalue μ if*

$$y^H A = \mu y^H. \tag{9.2}$$

By taking Hermitian transposes in (9.1), we see immediately that x^H is a left eigenvector of A^H associated with $\bar{\lambda}$. Note that if x $[y]$ is a right [left] eigenvector of A, then so is αx $[\alpha y]$ for any nonzero scalar $\alpha \in \mathbb{C}$. One often-used scaling for an eigenvector is $\alpha = 1/\|x\|$ so that the scaled eigenvector has norm 1. The 2-norm is the most common norm used for such scaling.

Definition 9.2. *The polynomial $\pi(\lambda) = \det(A - \lambda I)$ is called the* **characteristic polynomial** *of A. (Note that the characteristic polynomial can also be defined as $\det(\lambda I - A)$. This results in at most a change of sign and, as a matter of convenience, we use both forms throughout the text.)*

The following classical theorem can be very useful in hand calculation. It can be proved easily from the Jordan canonical form to be discussed in the text to follow (see, for example, [21]) or directly using elementary properties of inverses and determinants (see, for example, [3]).

Theorem 9.3 (Cayley–Hamilton). *For any $A \in \mathbb{C}^{n\times n}$, $\pi(A) = 0$.*

Example 9.4. Let $A = \begin{bmatrix} -7 & -4 \\ 8 & 5 \end{bmatrix}$. Then $\pi(\lambda) = \lambda^2 + 2\lambda - 3$. It is an easy exercise to verify that $\pi(A) = A^2 + 2A - 3I = 0$.

It can be proved from elementary properties of determinants that if $A \in \mathbb{C}^{n\times n}$, then $\pi(\lambda)$ is a polynomial of degree n. Thus, the Fundamental Theorem of Algebra says that

$\pi(\lambda)$ has n roots, possibly repeated. These roots, as solutions of the determinant equation

$$\pi(\lambda) = \det(A - \lambda I) = 0, \tag{9.3}$$

are the eigenvalues of A and imply the singularity of the matrix $A - \lambda I$, and hence further guarantee the existence of corresponding nonzero eigenvectors.

Definition 9.5. *The* **spectrum** *of $A \in \mathbb{C}^{n\times n}$ is the set of all eigenvalues of A, i.e., the set of all roots of its characteristic polynomial $\pi(\lambda)$. The spectrum of A is denoted $\Lambda(A)$.*

Let the eigenvalues of $A \in \mathbb{C}^{n\times n}$ be denoted $\lambda_1, \ldots, \lambda_n$. Then if we write (9.3) in the form

$$\pi(\lambda) = \det(A - \lambda I) = (\lambda_1 - \lambda)\cdots(\lambda_n - \lambda) \tag{9.4}$$

and set $\lambda = 0$ in this identity, we get the interesting fact that $\det(A) = \lambda_1 \cdot \lambda_2 \cdots \lambda_n$ (see also Theorem 9.25).

If $A \in \mathbb{R}^{n\times n}$, then $\pi(\lambda)$ has real coefficients. Hence the roots of $\pi(\lambda)$, i.e., the eigenvalues of A, must occur in complex conjugate pairs.

Example 9.6. Let $\alpha, \beta \in \mathbb{R}$ and let $A = \begin{bmatrix} \alpha & \beta \\ -\beta & \alpha \end{bmatrix}$. Then $\pi(\lambda) = \lambda^2 - 2\alpha\lambda + \alpha^2 + \beta^2$ and A has eigenvalues $\alpha \pm \beta j$ (where $j = i = \sqrt{-1}$).

If $A \in \mathbb{R}^{n\times n}$, then there is an easily checked relationship between the left and right eigenvectors of A and A^T (take Hermitian transposes of both sides of (9.2)). Specifically, if y is a left eigenvector of A corresponding to $\lambda \in \Lambda(A)$, then y is a right eigenvector of A^T corresponding to $\bar{\lambda} \in \Lambda(A)$. Note, too, that by elementary properties of the determinant, we always have $\Lambda(A) = \Lambda(A^T)$, but that $\overline{\Lambda(A)} = \Lambda(A)$ only if $A \in \mathbb{R}^{n\times n}$.

Definition 9.7. *If λ is a root of multiplicity m of $\pi(\lambda)$, we say that λ is an eigenvalue of A of* **algebraic multiplicity** *m. The* **geometric multiplicity** *of λ is the number of associated independent eigenvectors $= n - \operatorname{rank}(A - \lambda I) = \dim \mathcal{N}(A - \lambda I)$.*

If $\lambda \in \Lambda(A)$ has algebraic multiplicity m, then $1 \leq \dim \mathcal{N}(A - \lambda I) \leq m$. Thus, if we denote the geometric multiplicity of λ by g, then we must have $1 \leq g \leq m$.

Definition 9.8. *A matrix $A \in \mathbb{R}^{n\times n}$ is said to be* **defective** *if it has an eigenvalue whose geometric multiplicity is not equal to (i.e., less than) its algebraic multiplicity. Equivalently, A is said to be defective if it does not have n linearly independent (right or left) eigenvectors.*

From the Cayley–Hamilton Theorem, we know that $\pi(A) = 0$. However, it is possible for A to satisfy a lower-order polynomial. For example, if $A = \begin{bmatrix} 1 & 0 \\ 0 & 1 \end{bmatrix}$, then A satisfies $(\lambda - 1)^2 = 0$. But it also clearly satisfies the smaller degree polynomial equation $(\lambda - 1) = 0$.

Definition 9.9. *The* **minimal polynomial** *of $A \in \mathbb{R}^{n\times n}$ is the polynomial $\alpha(\lambda)$ of least degree such that $\alpha(A) = 0$.*

It can be shown that $\alpha(\lambda)$ is essentially unique (unique if we force the coefficient of the highest power of λ to be $+1$, say; such a polynomial is said to be **monic** and we generally write $\alpha(\lambda)$ as a monic polynomial throughout the text). Moreover, it can also be shown that $\alpha(\lambda)$ divides every nonzero polynomial $\beta(\lambda)$ for which $\beta(A) = 0$. In particular, $\alpha(\lambda)$ divides $\pi(\lambda)$.

There is an algorithm to determine $\alpha(\lambda)$ directly (without knowing eigenvalues and associated eigenvector structure). Unfortunately, this algorithm, called the Bezout algorithm, is numerically unstable.

Example 9.10. The above definitions are illustrated below for a series of matrices, each of which has an eigenvalue 2 of algebraic multiplicity 4, i.e., $\pi(\lambda) = (\lambda - 2)^4$. We denote the geometric multiplicity by g.

$$A = \begin{bmatrix} 2 & 1 & 0 & 0 \\ 0 & 2 & 1 & 0 \\ 0 & 0 & 2 & 1 \\ 0 & 0 & 0 & 2 \end{bmatrix} \text{ has } \alpha(\lambda) = (\lambda - 2)^4 \text{ and } g = 1.$$

$$A = \begin{bmatrix} 2 & 1 & 0 & 0 \\ 0 & 2 & 1 & 0 \\ 0 & 0 & 2 & 0 \\ 0 & 0 & 0 & 2 \end{bmatrix} \text{ has } \alpha(\lambda) = (\lambda - 2)^3 \text{ and } g = 2.$$

$$A = \begin{bmatrix} 2 & 1 & 0 & 0 \\ 0 & 2 & 0 & 0 \\ 0 & 0 & 2 & 0 \\ 0 & 0 & 0 & 2 \end{bmatrix} \text{ has } \alpha(\lambda) = (\lambda - 2)^2 \text{ and } g = 3.$$

$$A = \begin{bmatrix} 2 & 0 & 0 & 0 \\ 0 & 2 & 0 & 0 \\ 0 & 0 & 2 & 0 \\ 0 & 0 & 0 & 2 \end{bmatrix} \text{ has } \alpha(\lambda) = (\lambda - 2) \text{ and } g = 4.$$

At this point, one might speculate that g plus the degree of α must always be five. Unfortunately, such is not the case. The matrix

$$A = \begin{bmatrix} 2 & 1 & 0 & 0 \\ 0 & 2 & 0 & 0 \\ 0 & 0 & 2 & 1 \\ 0 & 0 & 0 & 2 \end{bmatrix}$$

has $\alpha(\lambda) = (\lambda - 2)^2$ and $g = 2$.

Theorem 9.11. *Let $A \in \mathbb{C}^{n \times n}$ and let λ_i be an eigenvalue of A with corresponding right eigenvector x_i. Furthermore, let y_j be a left eigenvector corresponding to any $\lambda_j \in \Lambda(A)$ such that $\lambda_j \neq \lambda_i$. Then $y_j^H x_i = 0$.*

Proof: Since $Ax_i = \lambda_i x_i$,

$$y_j^H A x_i = \lambda_i y_j^H x_i. \tag{9.5}$$

Similarly, since $y_j^H A = \lambda_j y_j^H$,

$$y_j^H A x_i = \lambda_j y_j^H x_i. \tag{9.6}$$

Subtracting (9.6) from (9.5), we find $0 = (\lambda_i - \lambda_j) y_j^H x_i$. Since $\lambda_i - \lambda_j \neq 0$, we must have $y_j^H x_i = 0$. □

The proof of Theorem 9.11 is very similar to two other fundamental and important results.

Theorem 9.12. *Let $A \in \mathbb{C}^{n \times n}$ be Hermitian, i.e., $A = A^H$. Then all eigenvalues of A must be real.*

Proof: Suppose (λ, x) is an arbitrary eigenvalue/eigenvector pair such that $Ax = \lambda x$. Then

$$x^H A x = \lambda x^H x. \tag{9.7}$$

Taking Hermitian transposes in (9.7) yields

$$x^H A^H x = \bar{\lambda} x^H x.$$

Using the fact that A is Hermitian, we have that $\bar{\lambda} x^H x = \lambda x^H x$. However, since x is an eigenvector, we have $x^H x \neq 0$, from which we conclude $\bar{\lambda} = \lambda$, i.e., λ is real. □

Theorem 9.13. *Let $A \in \mathbb{C}^{n \times n}$ be Hermitian and suppose λ and μ are distinct eigenvalues of A with corresponding right eigenvectors x and z, respectively. Then x and z must be orthogonal.*

Proof: Premultiply the equation $Ax = \lambda x$ by z^H to get $z^H A x = \lambda z^H x$. Take the Hermitian transpose of this equation and use the facts that A is Hermitian and λ is real to get $x^H A z = \lambda x^H z$. Premultiply the equation $Az = \mu z$ by x^H to get $x^H A z = \mu x^H z = \lambda x^H z$. Since $\lambda \neq \mu$, we must have that $x^H z = 0$, i.e., the two vectors must be orthogonal. □

Let us now return to the general case.

Theorem 9.14. *Let $A \in \mathbb{C}^{n \times n}$ have distinct eigenvalues $\lambda_1, \ldots, \lambda_n$ with corresponding right eigenvectors $x_1, \ldots, x_n$. Then $\{x_1, \ldots, x_n\}$ is a linearly independent set. The same result holds for the corresponding left eigenvectors.*

Proof: For the proof see, for example, [21, p. 118]. □

If $A \in \mathbb{C}^{n \times n}$ has distinct eigenvalues, and if $\lambda_i \in \Lambda(A)$, then by Theorem 9.11, x_i is orthogonal to all y_j's for which $j \neq i$. However, it cannot be the case that $y_i^H x_i = 0$ as well, or else x_i would be orthogonal to n linearly independent vectors (by Theorem 9.14) and would thus have to be 0, contradicting the fact that it is an eigenvector. Since $y_i^H x_i \neq 0$ for each i, we can choose the normalization of the x_i's, or the y_i's, or both, so that $y_i^H x_i = 1$ for $i \in \underline{n}$.

Theorem 9.15. *Let $A \in \mathbb{C}^{n\times n}$ have distinct eigenvalues $\lambda_1, \ldots, \lambda_n$ and let the corresponding right eigenvectors form a matrix $X = [x_1, \ldots, x_n]$. Similarly, let $Y = [y_1, \ldots, y_n]$ be the matrix of corresponding left eigenvectors. Furthermore, suppose that the left and right eigenvectors have been normalized so that $y_i^H x_i = 1$, $i \in \underline{n}$. Finally, let $\Lambda = \operatorname{diag}(\lambda_1, \ldots, \lambda_n) \in \mathbb{R}^{n\times n}$. Then $Ax_i = \lambda_i x_i$, $i \in \underline{n}$, can be written in matrix form as*

$$AX = X\Lambda \tag{9.8}$$

while $y_i^H x_j = \delta_{ij}$, $i \in \underline{n}$, $j \in \underline{n}$, is expressed by the equation

$$Y^H X = I. \tag{9.9}$$

These matrix equations can be combined to yield the following matrix factorizations:

$$X^{-1}AX = \Lambda = Y^H AX \tag{9.10}$$

and

$$A = X\Lambda X^{-1} = X\Lambda Y^H = \sum_{i=1}^{n} \lambda_i x_i y_i^H. \tag{9.11}$$

Example 9.16. Let

$$A = \begin{bmatrix} 2 & -3 & 1 \\ 5 & -2 & 5 \\ -3 & 1 & -4 \end{bmatrix}.$$

Then $\pi(\lambda) = \det(A - \lambda I) = -(\lambda^3 + 4\lambda^2 + 9\lambda + 10) = -(\lambda + 2)(\lambda^2 + 2\lambda + 5)$, from which we find $\Lambda(A) = \{-2, -1 \pm 2j\}$. We can now find the right and left eigenvectors corresponding to these eigenvalues.

For $\lambda_1 = -2$, solve the 3×3 linear system $(A - (-2)I)x_1 = 0$ to get

$$x_1 = \begin{bmatrix} -1 \\ -1 \\ 1 \end{bmatrix}.$$

Note that one component of x_1 can be set arbitrarily, and this then determines the other two (since $\dim \mathcal{N}(A - (-2)I) = 1$). To get the corresponding left eigenvector y_1, solve the linear system $y_1^T(A + 2I) = 0$ to get

$$y_1 = \begin{bmatrix} 1 \\ 1 \\ 3 \end{bmatrix}.$$

This time we have chosen the arbitrary scale factor for y_1 so that $y_1^T x_1 = 1$.

For $\lambda_2 = -1 + 2j$, solve the linear system $(A - (-1 + 2j)I)x_2 = 0$ to get

$$x_2 = \begin{bmatrix} 3 + j \\ 3 - j \\ -2 \end{bmatrix}.$$

Solve the linear system $y_2^H(A - (-1+2j)I) = 0$ and normalize y_2 so that $y_2^H x_2 = 1$ to get

$$y_2 = \frac{1}{4}\begin{bmatrix} 1+j \\ 1-j \\ 2 \end{bmatrix}.$$

For $\lambda_3 = -1 - 2j$, we could proceed to solve linear systems as for λ_2. However, we can also note that $x_3 = \overline{x_2}$ and $y_3 = \overline{y_2}$. To see this, use the fact that $\lambda_3 = \overline{\lambda_2}$ and simply conjugate the equation $Ax_2 = \lambda_2 x_2$ to get $A\overline{x_2} = \overline{\lambda_2}\,\overline{x_2}$. A similar argument yields the result for left eigenvectors.

Now define the matrix X of right eigenvectors:

$$X = [x_1 \quad x_2 \quad x_3] = \begin{bmatrix} -1 & 3+j & 3-j \\ -1 & 3-j & 3+j \\ 1 & -2 & -2 \end{bmatrix}.$$

It is then easy to verify that

$$X^{-1} = \begin{bmatrix} 1 & 1 & 3 \\ \frac{1-j}{4} & \frac{1+j}{4} & \frac{1}{2} \\ \frac{1+j}{4} & \frac{1-j}{4} & \frac{1}{2} \end{bmatrix} = Y^H = \begin{bmatrix} y_1^H \\ y_2^H \\ y_3^H \end{bmatrix}.$$

Other results in Theorem 9.15 can also be verified. For example,

$$X^{-1}AX = \Lambda = \begin{bmatrix} -2 & 0 & 0 \\ 0 & -1+2j & 0 \\ 0 & 0 & -1-2j \end{bmatrix}.$$

Finally, note that we could have solved directly only for x_1 and x_2 (and $x_3 = \overline{x_2}$). Then, instead of determining the y_i's directly, we could have found them instead by computing X^{-1} and reading off its rows.

Example 9.17. Let

$$A = \begin{bmatrix} -3 & 1 & 0 \\ 1 & -2 & 1 \\ 0 & 1 & -3 \end{bmatrix}.$$

Then $\pi(\lambda) = \det(A - \lambda I) = -(\lambda^3 + 8\lambda^2 + 19\lambda + 12) = -(\lambda+1)(\lambda+3)(\lambda+4)$, from which we find $\Lambda(A) = \{-1, -3, -4\}$. Proceeding as in the previous example, it is straightforward to compute

$$X = \begin{bmatrix} 1 & 1 & 1 \\ 2 & 0 & -1 \\ 1 & -1 & 1 \end{bmatrix}$$

and

$$X^{-1} = \frac{1}{6}\begin{bmatrix} 1 & 2 & 1 \\ 3 & 0 & -3 \\ 2 & -2 & 2 \end{bmatrix} = Y^T.$$

We also have $X^{-1}AX = \Lambda = \text{diag}(-1, -3, -4)$, which is equivalent to the dyadic expansion

$$A = \sum_{i=1}^{3} \lambda_i x_i y_i^T$$

$$= (-1)\begin{bmatrix} 1 \\ 2 \\ 1 \end{bmatrix}\begin{bmatrix} \frac{1}{6} & \frac{1}{3} & \frac{1}{6} \end{bmatrix} + (-3)\begin{bmatrix} 1 \\ 0 \\ -1 \end{bmatrix}\begin{bmatrix} \frac{1}{2} & 0 & -\frac{1}{2} \end{bmatrix}$$

$$+ (-4)\begin{bmatrix} 1 \\ -1 \\ 1 \end{bmatrix}\begin{bmatrix} \frac{1}{3} & -\frac{1}{3} & \frac{1}{3} \end{bmatrix}$$

$$= (-1)\begin{bmatrix} \frac{1}{6} & \frac{1}{3} & \frac{1}{6} \\ \frac{1}{3} & \frac{2}{3} & \frac{1}{3} \\ \frac{1}{6} & \frac{1}{3} & \frac{1}{6} \end{bmatrix} + (-3)\begin{bmatrix} \frac{1}{2} & 0 & -\frac{1}{2} \\ 0 & 0 & 0 \\ -\frac{1}{2} & 0 & \frac{1}{2} \end{bmatrix} + (-4)\begin{bmatrix} \frac{1}{3} & -\frac{1}{3} & \frac{1}{3} \\ -\frac{1}{3} & \frac{1}{3} & -\frac{1}{3} \\ \frac{1}{3} & -\frac{1}{3} & \frac{1}{3} \end{bmatrix}.$$

Theorem 9.18. *Eigenvalues (but not eigenvectors) are invariant under a similarity transformation T.*

Proof: Suppose (λ, x) is an eigenvalue/eigenvector pair such that $Ax = \lambda x$. Then, since T is nonsingular, we have the equivalent statement $(T^{-1}AT)(T^{-1}x) = \lambda(T^{-1}x)$, from which the theorem statement follows. For left eigenvectors we have a similar statement, namely $y^H A = \lambda y^H$ if and only if $(T^H y)^H (T^{-1}AT) = \lambda (T^H y)^H$. $\square$

Remark 9.19. If f is an analytic function (e.g., $f(x)$ is a polynomial, or e^x, or $\sin x$, or, in general, representable by a power series $\sum_{n=0}^{+\infty} a_n x^n$), then it is easy to show that the eigenvalues of $f(A)$ (defined as $\sum_{n=0}^{+\infty} a_n A^n$) are $f(\lambda)$, but $f(A)$ does not necessarily have all the same eigenvectors (unless, say, A is diagonalizable). For example, $A = \begin{bmatrix} 0 & 1 \\ 0 & 0 \end{bmatrix}$ has only one right eigenvector corresponding to the eigenvalue 0, but $A^2 = \begin{bmatrix} 0 & 0 \\ 0 & 0 \end{bmatrix}$ has two independent right eigenvectors associated with the eigenvalue 0. What is true is that the eigenvalue/eigenvector pair (λ, x) maps to $(f(\lambda), x)$ but not conversely.

The following theorem is useful when solving systems of linear differential equations. Details of how the matrix exponential e^{tA} is used to solve the system $\dot{x} = Ax$ are the subject of Chapter 11.

Theorem 9.20. *Let $A \in \mathbb{R}^{n \times n}$ and suppose $X^{-1}AX = \Lambda$, where Λ is diagonal. Then*

$$e^{tA} = X\text{diag}(e^{\lambda_1 t}, \ldots, e^{\lambda_n t})X^{-1}$$

$$= \sum_{i=1}^{n} e^{\lambda_i t} x_i y_i^H.$$

Proof: Starting from the definition, we have

$$\begin{aligned}
e^{tA} &= \sum_{k=0}^{+\infty} \frac{t^k}{k!} A^k \\
&= \sum_{k=0}^{+\infty} \frac{t^k}{k!} (X \Lambda X^{-1})^k \\
&= \sum_{k=0}^{+\infty} \frac{t^k}{k!} X \Lambda^k X^{-1} \\
&= X \left(\sum_{k=0}^{+\infty} \frac{t^k}{k!} \Lambda^k \right) X^{-1} \\
&= X e^{t\Lambda} X^{-1} \\
&= X \mathrm{diag}(e^{\lambda_1 t}, \ldots, e^{\lambda_n t}) Y^H \\
&= \sum_{i=1}^{n} e^{\lambda_i t} x_i y_i^H. \qquad \square
\end{aligned}$$

The following corollary is immediate from the theorem upon setting $t = 1$.

Corollary 9.21. *If $A \in \mathbb{R}^{n \times n}$ is diagonalizable with eigenvalues λ_i, $i \in \underline{n}$, and right eigenvectors x_i, $i \in \underline{n}$, then e^A has eigenvalues e^{λ_i}, $i \in \underline{n}$, and the same eigenvectors.*

There are extensions to Theorem 9.20 and Corollary 9.21 for any function that is analytic on the spectrum of A, i.e., $f(A) = X f(\Lambda) X^{-1} = X \mathrm{diag}(f(\lambda_1), \ldots, f(\lambda_n)) X^{-1}$.

It is desirable, of course, to have a version of Theorem 9.20 and its corollary in which A is not necessarily diagonalizable. It is necessary first to consider the notion of Jordan canonical form, from which such a result is then available and presented later in this chapter.

9.2 Jordan Canonical Form

Theorem 9.22.

1. *Jordan Canonical Form (JCF): For all $A \in \mathbb{C}^{n \times n}$ with eigenvalues $\lambda_1, \ldots, \lambda_n \in \mathbb{C}$ (not necessarily distinct), there exists $X \in \mathbb{C}_n^{n \times n}$ such that*

$$X^{-1} A X = J = \mathrm{diag}(J_1, \ldots, J_q), \tag{9.12}$$

where each of the Jordan block matrices $J_1, \ldots, J_q$ is of the form

$$J_i = \begin{bmatrix} \lambda_i & 1 & 0 & \cdots & & 0 \\ 0 & \lambda_i & 1 & 0 & & \vdots \\ \vdots & \ddots & \lambda_i & \ddots & \ddots & \\ & & & \ddots & 1 & 0 \\ \vdots & & & \ddots & \lambda_i & 1 \\ 0 & \cdots & & \cdots & 0 & \lambda_i \end{bmatrix} \in \mathbb{C}^{k_i \times k_i} \tag{9.13}$$

and $\sum_{i=1}^{q} k_i = n$.

2. *Real Jordan Canonical Form: For all* $A \in \mathbb{R}^{n \times n}$ *with eigenvalues* $\lambda_1, \ldots, \lambda_n$ *(not necessarily distinct), there exists* $X \in \mathbb{R}_n^{n \times n}$ *such that*

$$X^{-1}AX = J = \operatorname{diag}(J_1, \ldots, J_q), \tag{9.14}$$

where each of the Jordan block matrices $J_1, \ldots, J_q$ *is of the form*

$$J_i = \begin{bmatrix} \lambda_i & 1 & 0 & \cdots & & 0 \\ 0 & \lambda_i & 1 & 0 & & \vdots \\ \vdots & \ddots & \lambda_i & \ddots & \ddots & \\ & & & \ddots & 1 & 0 \\ \vdots & & & \ddots & \lambda_i & 1 \\ 0 & \cdots & & \cdots & 0 & \lambda_i \end{bmatrix} \tag{9.15}$$

in the case of real eigenvalues $\lambda_i \in \Lambda(A)$, *and*

$$J_i = \begin{bmatrix} M_i & I_2 & 0 & \cdots & & 0 \\ 0 & M_i & I_2 & 0 & & \vdots \\ \vdots & \ddots & M_i & \ddots & \ddots & \\ & & & \ddots & I_2 & 0 \\ \vdots & & & \ddots & M_i & I_2 \\ 0 & \cdots & & \cdots & 0 & M_i \end{bmatrix}, \tag{9.16}$$

where $M_i = \left[\begin{smallmatrix} \alpha_i & \beta_i \\ -\beta_i & \alpha_i \end{smallmatrix}\right]$ *and* $I_2 = \left[\begin{smallmatrix} 1 & 0 \\ 0 & 1 \end{smallmatrix}\right]$ *in the case of complex conjugate eigenvalues* $\alpha_i \pm j\beta_i \in \Lambda(A)$.

Proof: For the proof see, for example, [21, pp. 120–124]. $\square$

Transformations like $T = \left[\begin{smallmatrix} 1 & -j \\ -j & 1 \end{smallmatrix}\right]$ allow us to go back and forth between a real JCF and its complex counterpart:

$$T^{-1} \begin{bmatrix} \alpha + j\beta & 0 \\ 0 & \alpha - j\beta \end{bmatrix} T = \begin{bmatrix} \alpha & \beta \\ -\beta & \alpha \end{bmatrix} = M.$$

For nontrivial Jordan blocks, the situation is only a bit more complicated. With

$$T = \begin{bmatrix} 1 & -j & 0 & 0 \\ 0 & 0 & 1 & -j \\ -j & 1 & 0 & 0 \\ 0 & 0 & -j & 1 \end{bmatrix},$$

it is easily checked that

$$T^{-1}\begin{bmatrix} \alpha+j\beta & 1 & 0 & 0 \\ 0 & \alpha+j\beta & 0 & 0 \\ 0 & 0 & \alpha-j\beta & 1 \\ 0 & 0 & 0 & \alpha-j\beta \end{bmatrix}T = \begin{bmatrix} M & I_2 \\ 0 & M \end{bmatrix}.$$

Definition 9.23. *The characteristic polynomials of the Jordan blocks defined in Theorem 9.22 are called the* **elementary divisors** *or* **invariant factors** *of* A.

Theorem 9.24. *The characteristic polynomial of a matrix is the product of its elementary divisors. The minimal polynomial of a matrix is the product of the elementary divisors of highest degree corresponding to distinct eigenvalues.*

Theorem 9.25. *Let* $A \in \mathbb{C}^{n\times n}$ *with eigenvalues* $\lambda_1, \ldots, \lambda_n$. *Then*

1. $\det(A) = \prod_{i=1}^{n} \lambda_i$.

2. $\operatorname{Tr}(A) = \sum_{i=1}^{n} \lambda_i$.

Proof:

1. From Theorem 9.22 we have that $A = XJX^{-1}$. Thus, $\det(A) = \det(XJX^{-1}) = \det(J) = \prod_{i=1}^{n} \lambda_i$.

2. Again, from Theorem 9.22 we have that $A = XJX^{-1}$. Thus, $\operatorname{Tr}(A) = \operatorname{Tr}(XJX^{-1}) = \operatorname{Tr}(JX^{-1}X) = \operatorname{Tr}(J) = \sum_{i=1}^{n} \lambda_i$. □

Example 9.26. Suppose $A \in \mathbb{R}^{7\times 7}$ is known to have $\pi(\lambda) = (\lambda-1)^4(\lambda-2)^3$ and $\alpha(\lambda) = (\lambda-1)^2(\lambda-2)^2$. Then A has two possible JCFs (not counting reorderings of the diagonal blocks):

$$J^{(1)} = \begin{bmatrix} 1 & 1 & 0 & 0 & 0 & 0 & 0 \\ 0 & 1 & 0 & 0 & 0 & 0 & 0 \\ 0 & 0 & 1 & 0 & 0 & 0 & 0 \\ 0 & 0 & 0 & 1 & 0 & 0 & 0 \\ 0 & 0 & 0 & 0 & 2 & 1 & 0 \\ 0 & 0 & 0 & 0 & 0 & 2 & 0 \\ 0 & 0 & 0 & 0 & 0 & 0 & 2 \end{bmatrix} \quad \text{and} \quad J^{(2)} = \begin{bmatrix} 1 & 1 & 0 & 0 & 0 & 0 & 0 \\ 0 & 1 & 0 & 0 & 0 & 0 & 0 \\ 0 & 0 & 1 & 1 & 0 & 0 & 0 \\ 0 & 0 & 0 & 1 & 0 & 0 & 0 \\ 0 & 0 & 0 & 0 & 2 & 1 & 0 \\ 0 & 0 & 0 & 0 & 0 & 2 & 0 \\ 0 & 0 & 0 & 0 & 0 & 0 & 2 \end{bmatrix}.$$

Note that $J^{(1)}$ has elementary divisors $(\lambda-1)^2$, $(\lambda-1)$, $(\lambda-1)$, $(\lambda-2)^2$, and $(\lambda-2)$, while $J^{(2)}$ has elementary divisors $(\lambda-1)^2$, $(\lambda-1)^2$, $(\lambda-2)^2$, and $(\lambda-2)$.

Example 9.27. Knowing $\pi(\lambda)$, $\alpha(\lambda)$, and $\mathrm{rank}(A - \lambda_i I)$ for distinct λ_i is **not** sufficient to determine the JCF of A uniquely. The matrices

$$A_1 = \begin{bmatrix} a & 1 & 0 & 0 & 0 & 0 & 0 \\ 0 & a & 1 & 0 & 0 & 0 & 0 \\ 0 & 0 & a & 0 & 0 & 0 & 0 \\ 0 & 0 & 0 & a & 1 & 0 & 0 \\ 0 & 0 & 0 & 0 & a & 0 & 0 \\ 0 & 0 & 0 & 0 & 0 & a & 1 \\ 0 & 0 & 0 & 0 & 0 & 0 & a \end{bmatrix}, \quad A_2 = \begin{bmatrix} a & 1 & 0 & 0 & 0 & 0 & 0 \\ 0 & a & 1 & 0 & 0 & 0 & 0 \\ 0 & 0 & a & 0 & 0 & 0 & 0 \\ 0 & 0 & 0 & a & 1 & 0 & 0 \\ 0 & 0 & 0 & 0 & a & 1 & 0 \\ 0 & 0 & 0 & 0 & 0 & a & 0 \\ 0 & 0 & 0 & 0 & 0 & 0 & a \end{bmatrix}$$

both have $\pi(\lambda) = (\lambda - a)^7$, $\alpha(\lambda) = (\lambda - a)^3$, and $\mathrm{rank}(A - aI) = 4$, i.e., three eigenvectors.

9.3 Determination of the JCF

The first critical item of information in determining the JCF of a matrix $A \in \mathbb{R}^{n \times n}$ is its number of eigenvectors. For each distinct eigenvalue λ_i, the associated number of linearly independent right (or left) eigenvectors is given by $\dim \mathcal{N}(A - \lambda_i I) = n - \mathrm{rank}(A - \lambda_i I)$. The straightforward case is, of course, when λ_i is **simple**, i.e., of algebraic multiplicity 1; it then has precisely one eigenvector. The more interesting (and difficult) case occurs when λ_i is of algebraic multiplicity greater than one. For example, suppose

$$A = \begin{bmatrix} 3 & 2 & 1 \\ 0 & 3 & 0 \\ 0 & 0 & 3 \end{bmatrix}.$$

Then

$$A - 3I = \begin{bmatrix} 0 & 2 & 1 \\ 0 & 0 & 0 \\ 0 & 0 & 0 \end{bmatrix}$$

has rank 1, so the eigenvalue 3 has two eigenvectors associated with it. If we let $[\xi_1 \ \ \xi_2 \ \ \xi_3]^T$ denote a solution to the linear system $(A - 3I)\xi = 0$, we find that $2\xi_2 + \xi_3 = 0$. Thus, both

$$\xi = x_1 = \begin{bmatrix} 1 \\ 0 \\ 0 \end{bmatrix} \text{ and } \xi = x_2 = \begin{bmatrix} 0 \\ 1 \\ -2 \end{bmatrix}$$

are eigenvectors (and are independent). To get a third vector x_3 such that $X = [x_1 \ \ x_2 \ \ x_3]$ reduces A to JCF, we need the notion of principal vector.

Definition 9.28. *Let $A \in \mathbb{C}^{n \times n}$ (or $\mathbb{R}^{n \times n}$). Then x is a* **right principal vector of degree** k *associated with $\lambda \in \Lambda(A)$ if and only if $(A - \lambda I)^k x = 0$ and $(A - \lambda I)^{k-1} x \neq 0$.*

Remark 9.29.

1. An analogous definition holds for a left principal vector of degree k.

2. The phrase "of grade k" is often used synonymously with "of degree k."
3. Principal vectors are sometimes also called **generalized eigenvectors**, but the latter term will be assigned a much different meaning in Chapter 12.
4. The case $k = 1$ corresponds to the "usual" eigenvector.
5. A right (or left) principal vector of degree k is associated with a Jordan block J_i of dimension k or larger.

9.3.1 Theoretical computation

To motivate the development of a procedure for determining principal vectors, consider a 2×2 Jordan block $\left[\begin{smallmatrix} \lambda & 1 \\ 0 & \lambda \end{smallmatrix}\right]$. Denote by $x^{(1)}$ and $x^{(2)}$ the two columns of a matrix $X \in \mathbb{R}_2^{2\times 2}$ that reduces a matrix A to this JCF. Then the equation $AX = XJ$ can be written

$$A\begin{bmatrix} x^{(1)} & x^{(2)} \end{bmatrix} = \begin{bmatrix} x^{(1)} & x^{(2)} \end{bmatrix}\begin{bmatrix} \lambda & 1 \\ 0 & \lambda \end{bmatrix}.$$

The first column yields the equation $Ax^{(1)} = \lambda x^{(1)}$, which simply says that $x^{(1)}$ is a right eigenvector. The second column yields the following equation for $x^{(2)}$, the principal vector of degree 2:

$$(A - \lambda I)x^{(2)} = x^{(1)}. \tag{9.17}$$

If we premultiply (9.17) by $(A - \lambda I)$, we find $(A - \lambda I)^2 x^{(2)} = (A - \lambda I)x^{(1)} = 0$. Thus, the definition of principal vector is satisfied.

This suggests a "general" procedure. First, determine all eigenvalues of $A \in \mathbb{R}^{n\times n}$ (or $\mathbb{C}^{n\times n}$). Then for each distinct $\lambda \in \Lambda(A)$ perform the following:

1. Solve

   $$(A - \lambda I)x^{(1)} = 0.$$

   This step finds all the eigenvectors (i.e., principal vectors of degree 1) associated with λ. The number of eigenvectors depends on the rank of $A - \lambda I$. For example, if $\operatorname{rank}(A - \lambda I) = n - 1$, there is only one eigenvector. If the algebraic multiplicity of λ is greater than its geometric multiplicity, principal vectors still need to be computed from succeeding steps.

2. For each independent $x^{(1)}$, solve

   $$(A - \lambda I)x^{(2)} = x^{(1)}.$$

   The number of linearly independent solutions at this step depends on the rank of $(A - \lambda I)^2$. If, for example, this rank is $n - 2$, there are two linearly independent solutions to the homogeneous equation $(A - \lambda I)^2 x^{(2)} = 0$. One of these solutions is, of course, $x^{(1)}$ $(\neq 0)$, since $(A - \lambda I)^2 x^{(1)} = (A - \lambda I)0 = 0$. The other solution is the desired principal vector of degree 2. (It may be necessary to take a linear combination of $x^{(1)}$ vectors to get a right-hand side that is in $\mathcal{R}(A - \lambda I)$. See, for example, Exercise 7.)

3. For each independent $x^{(2)}$ from step 2, solve

$$(A - \lambda I)x^{(3)} = x^{(2)}.$$

4. Continue in this way until the total number of independent eigenvectors and principal vectors is equal to the algebraic multiplicity of λ.

Unfortunately, this natural-looking procedure can fail to find all Jordan vectors. For more extensive treatments, see, for example, [20] and [21]. Determination of eigenvectors and principal vectors is obviously very tedious for anything beyond simple problems ($n = 2$ or 3, say). Attempts to do such calculations in finite-precision floating-point arithmetic generally prove unreliable. There are significant numerical difficulties inherent in attempting to compute a JCF, and the interested student is strongly urged to consult the classical and very readable [8] to learn why. Notice that high-quality mathematical software such as MATLAB does not offer a `jcf` command, although a `jordan` command is available in MATLAB's Symbolic Toolbox.

Theorem 9.30. *Suppose $A \in \mathbb{C}^{k \times k}$ has an eigenvalue λ of algebraic multiplicity k and suppose further that* $\operatorname{rank}(A - \lambda I) = k - 1$. *Let $X = \left[x^{(1)}, \ldots, x^{(k)}\right]$, where the* **chain** *of vectors $x^{(i)}$ is constructed as above. Then*

$$X^{-1}AX = \begin{bmatrix} \lambda & 1 & 0 & \cdots & & 0 \\ 0 & \lambda & 1 & 0 & & \vdots \\ \vdots & \ddots & \lambda & \ddots & \ddots & \\ & & & \ddots & 1 & 0 \\ \vdots & & & \ddots & \lambda & 1 \\ 0 & \cdots & & \cdots & 0 & \lambda \end{bmatrix}.$$

Theorem 9.31. $\{x^{(1)}, \ldots, x^{(k)}\}$ *is a linearly independent set.*

Theorem 9.32. *Principal vectors associated with different Jordan blocks are linearly independent.*

Example 9.33. Let

$$A = \begin{bmatrix} 1 & 1 & 2 \\ 0 & 1 & 3 \\ 0 & 0 & 2 \end{bmatrix}.$$

The eigenvalues of A are $\lambda_1 = 1$, $\lambda_2 = 1$, and $\lambda_3 = 2$. First, find the eigenvectors associated with the distinct eigenvalues 1 and 2.
$(A - 2I)x_3^{(1)} = 0$ yields

$$x_3^{(1)} = \begin{bmatrix} 5 \\ 3 \\ 1 \end{bmatrix}.$$

$(A - 1I)x_1^{(1)} = 0$ yields

$$x_1^{(1)} = \begin{bmatrix} 1 \\ 0 \\ 0 \end{bmatrix}.$$

To find a principal vector of degree 2 associated with the multiple eigenvalue 1, solve $(A - 1I)x_1^{(2)} = x_1^{(1)}$ to get

$$x_1^{(2)} = \begin{bmatrix} 0 \\ 1 \\ 0 \end{bmatrix}.$$

Now let

$$X = \begin{bmatrix} x_1^{(1)} & x_1^{(2)} & x_3^{(1)} \end{bmatrix} = \begin{bmatrix} 1 & 0 & 5 \\ 0 & 1 & 3 \\ 0 & 0 & 1 \end{bmatrix}.$$

Then it is easy to check that

$$X^{-1} = \begin{bmatrix} 1 & 0 & -5 \\ 0 & 1 & -3 \\ 0 & 0 & 1 \end{bmatrix} \text{ and } X^{-1}AX = \begin{bmatrix} 1 & 1 & 0 \\ 0 & 1 & 0 \\ 0 & 0 & 2 \end{bmatrix}.$$

9.3.2 On the +1's in JCF blocks

In this subsection we show that the nonzero superdiagonal elements of a JCF need not be 1's but can be arbitrary — so long as they are nonzero. For the sake of definiteness, we consider below the case of a single Jordan block, but the result clearly holds for any JCF. Suppose $A \in \mathbb{R}^{n \times n}$ and

$$X^{-1}AX = J = \begin{bmatrix} \lambda & 1 & 0 & \cdots & & 0 \\ 0 & \lambda & 1 & 0 & & \vdots \\ \vdots & \ddots & \lambda & \ddots & \ddots & \\ & & & \ddots & 1 & 0 \\ \vdots & & & \ddots & \lambda & 1 \\ 0 & \cdots & & \cdots & 0 & \lambda \end{bmatrix}.$$

Let $D = \text{diag}(d_1, \ldots, d_n)$ be a nonsingular "scaling" matrix. Then

$$D^{-1}(X^{-1}AX)D = D^{-1}JD = \hat{J} = \begin{bmatrix} \lambda & \frac{d_2}{d_1} & 0 & \cdots & & 0 \\ 0 & \lambda & \frac{d_3}{d_2} & 0 & & \vdots \\ \vdots & \ddots & \lambda & \ddots & \ddots & \\ & & & \ddots & \frac{d_{n-1}}{d_{n-2}} & 0 \\ \vdots & & & \ddots & \lambda & \frac{d_n}{d_{n-1}} \\ 0 & \cdots & & \cdots & 0 & \lambda \end{bmatrix}.$$

Appropriate choice of the d_i's then yields any desired nonzero superdiagonal elements. This result can also be interpreted in terms of the matrix $X = [x_1, \ldots, x_n]$ of eigenvectors and principal vectors that reduces A to its JCF. Specifically, $\hat{J}$ is obtained from A via the similarity transformation $XD = [d_1x_1, \ldots, d_nx_n]$.

In a similar fashion, the **reverse-order identity matrix** (or exchange matrix)

$$P = P^T = P^{-1} = \begin{bmatrix} 0 & \cdots & \cdots & 0 & 1 \\ \vdots & & \iddots & 1 & 0 \\ \vdots & \iddots & \iddots & \iddots & \vdots \\ 0 & 1 & \iddots & & \vdots \\ 1 & 0 & \cdots & \cdots & 0 \end{bmatrix} \tag{9.18}$$

can be used to put the superdiagonal elements in the subdiagonal instead if that is desired:

$$P^{-1} \begin{bmatrix} \lambda & 1 & 0 & \cdots & & 0 \\ 0 & \lambda & 1 & 0 & & \vdots \\ \vdots & \ddots & \lambda & \ddots & \ddots & \\ & & & \ddots & 1 & 0 \\ \vdots & & & \ddots & \lambda & 1 \\ 0 & \cdots & & \cdots & 0 & \lambda \end{bmatrix} P = \begin{bmatrix} \lambda & 0 & \cdots & & \cdots & 0 \\ 1 & \lambda & 0 & & & \vdots \\ 0 & 1 & \lambda & \ddots & & \\ \vdots & \ddots & \ddots & \ddots & \ddots & \vdots \\ & & \ddots & 1 & \lambda & 0 \\ 0 & \cdots & & 0 & 1 & \lambda \end{bmatrix}.$$

9.4 Geometric Aspects of the JCF

The matrix X that reduces a matrix $A \in \mathbb{R}^{n \times n}$ (or $\mathbb{C}^{n \times n}$) to a JCF provides a change of basis with respect to which the matrix is diagonal or block diagonal. It is thus natural to expect an associated direct sum decomposition of $\mathbb{R}^n$. Such a decomposition is given in the following theorem.

Theorem 9.34. *Suppose $A \in \mathbb{R}^{n \times n}$ has characteristic polynomial*

$$\pi(\lambda) = (\lambda - \lambda_1)^{n_1} \cdots (\lambda - \lambda_m)^{n_m}$$

and minimal polynomial

$$\alpha(\lambda) = (\lambda - \lambda_1)^{\nu_1} \cdots (\lambda - \lambda_m)^{\nu_m}$$

with $\lambda_1, \ldots, \lambda_m$ distinct. Then

$$\begin{aligned} \mathbb{R}^n &= \mathcal{N}(A - \lambda_1 I)^{n_1} \oplus \cdots \oplus \mathcal{N}(A - \lambda_m I)^{n_m} \\ &= \mathcal{N}(A - \lambda_1 I)^{\nu_1} \oplus \cdots \oplus \mathcal{N}(A - \lambda_m I)^{\nu_m}. \end{aligned}$$

Note that $\dim \mathcal{N}(A - \lambda_i I)^{\nu_i} = n_i$.

Definition 9.35. *Let $\mathcal{V}$ be a vector space over $\mathbb{F}$ and suppose $A : \mathcal{V} \to \mathcal{V}$ is a linear transformation. A subspace $\mathcal{S} \subseteq \mathcal{V}$ is* **A-invariant** *if $A\mathcal{S} \subseteq \mathcal{S}$, where $A\mathcal{S}$ is defined as the set $\{As : s \in \mathcal{S}\}$.*

If $\mathcal{V}$ is taken to be $\mathbb{R}^n$ over $\mathbb{R}$, and $S \in \mathbb{R}^{n \times k}$ is a matrix whose columns $s_1, \ldots, s_k$ span a k-dimensional subspace $\mathcal{S}$, i.e., $\mathcal{R}(S) = \mathcal{S}$, then $\mathcal{S}$ is A-invariant if and only if there exists $M \in \mathbb{R}^{k \times k}$ such that

$$AS = SM. \tag{9.19}$$

This follows easily by comparing the ith columns of each side of (9.19):

$$As_i = m_{1i}s_1 + m_{2i}s_2 + \cdots + m_{ki}s_k \in \mathcal{S}.$$

Example 9.36. The equation $Ax = \lambda x = x\lambda$ defining a right eigenvector x of an eigenvalue λ says that x spans an A-invariant subspace (of dimension one).

Example 9.37. Suppose X block diagonalizes A, i.e.,

$$X^{-1}AX = \begin{bmatrix} J_1 & 0 \\ 0 & J_2 \end{bmatrix}.$$

Rewriting in the form

$$A[X_1 \quad X_2] = [X_1 \quad X_2]\begin{bmatrix} J_1 & 0 \\ 0 & J_2 \end{bmatrix},$$

we have that $AX_i = X_iJ_i$, $i = 1, 2$, so the columns of X_i span an A-invariant subspace.

Theorem 9.38. *Suppose $A \in \mathbb{R}^{n \times n}$.*

1. *Let $p(A) = \alpha_0 I + \alpha_1 A + \cdots + \alpha_q A^q$ be a polynomial in A. Then $\mathcal{N}(p(A))$ and $\mathcal{R}(p(A))$ are A-invariant.*

2. *$\mathcal{S}$ is A-invariant if and only if $\mathcal{S}^\perp$ is A^T-invariant.*

Theorem 9.39. *If $\mathcal{V}$ is a vector space over $\mathbb{F}$ such that $\mathcal{V} = \mathcal{N}_1 \oplus \cdots \oplus \mathcal{N}_m$, where each $\mathcal{N}_i$ is A-invariant, then a basis for $\mathcal{V}$ can be chosen with respect to which A has a block diagonal representation.*

The Jordan canonical form is a special case of the above theorem. If A has distinct eigenvalues λ_i as in Theorem 9.34, we could choose bases for $\mathcal{N}(A - \lambda_i I)^{n_i}$ by SVD, for example (note that the power n_i could be replaced by ν_i). We would then get a block diagonal representation for A with full blocks rather than the highly structured Jordan blocks. Other such "canonical" forms are discussed in text that follows.

Suppose $X = [X_1, \ldots, X_m] \in \mathbb{R}_n^{n \times n}$ is such that $X^{-1}AX = \operatorname{diag}(J_1, \ldots, J_m)$, where each $J_i = \operatorname{diag}(J_{i1}, \ldots, J_{ik_i})$ and each J_{ik} is a Jordan block corresponding to $\lambda_i \in \Lambda(A)$. We could also use other block diagonal decompositions (e.g., via SVD), but we restrict our attention here to only the Jordan block case. Note that $AX_i = X_iJ_i$, so by (9.19) the columns of X_i (i.e., the eigenvectors and principal vectors associated with λ_i) span an A-invariant subspace of $\mathbb{R}^n$.

Finally, we return to the problem of developing a formula for e^{tA} in the case that A is not necessarily diagonalizable. Let $Y_i \in \mathbb{C}^{n \times n_i}$ be a Jordan basis for $\mathcal{N}(A^T - \overline{\lambda}_i I)^{n_i}$. Equivalently, partition

$$X^{-1} = \begin{bmatrix} Y_1^H \\ \vdots \\ Y_m^H \end{bmatrix}$$

compatibly. Then

$$\begin{aligned} A &= XJX^{-1} = XJY^H \\ &= [X_1, \ldots, X_m] \operatorname{diag}(J_1, \ldots, J_m) [Y_1, \ldots, Y_m]^H \\ &= \sum_{i=1}^{m} X_i J_i Y_i^H. \end{aligned}$$

In a similar fashion we can compute

$$e^{tA} = \sum_{i=1}^{m} X_i e^{tJ_i} Y_i^H,$$

which is a useful formula when used in conjunction with the result

$$\exp t \begin{bmatrix} \lambda & 1 & 0 & \cdots & 0 \\ 0 & \lambda & 1 & \ddots & \vdots \\ \vdots & \ddots & \lambda & \ddots & 0 \\ & & \ddots & \ddots & 1 \\ 0 & \cdots & & 0 & \lambda \end{bmatrix} = \begin{bmatrix} e^{\lambda t} & te^{\lambda t} & \frac{1}{2!}t^2 e^{\lambda t} & \cdots & \frac{1}{(k-1)!}t^{k-1}e^{\lambda t} \\ 0 & e^{\lambda t} & te^{\lambda t} & \ddots & \vdots \\ 0 & 0 & e^{\lambda t} & \ddots & \\ \vdots & & \ddots & \ddots & te^{\lambda t} \\ 0 & \cdots & & 0 & e^{\lambda t} \end{bmatrix}$$

for a $k \times k$ Jordan block J_i associated with an eigenvalue $\lambda = \lambda_i$.

9.5 The Matrix Sign Function

In this section we give a very brief introduction to an interesting and useful matrix function called the **matrix sign function**. It is a generalization of the sign (or signum) of a scalar. A survey of the matrix sign function and some of its applications can be found in [15].

Definition 9.40. *Let $z \in \mathbb{C}$ with $\operatorname{Re}(z) \neq 0$. Then the sign of z is defined by*

$$\operatorname{sgn}(z) = \frac{\operatorname{Re}(z)}{|\operatorname{Re}(z)|} = \begin{cases} +1 & \text{if } \operatorname{Re}(z) > 0, \\ -1 & \text{if } \operatorname{Re}(z) < 0. \end{cases}$$

Definition 9.41. *Suppose $A \in \mathbb{C}^{n \times n}$ has no eigenvalues on the imaginary axis, and let*

$$X^{-1}AX = \begin{bmatrix} N & 0 \\ 0 & P \end{bmatrix}$$

be a Jordan canonical form for A, with N containing all Jordan blocks corresponding to the eigenvalues of A in the left half-plane and P containing all Jordan blocks corresponding to eigenvalues in the right half-plane. Then the **sign** *of A, denoted $\operatorname{sgn}(A)$, is given by*

$$\operatorname{sgn}(A) = X \begin{bmatrix} -I & 0 \\ 0 & I \end{bmatrix} X^{-1},$$

where the negative and positive identity matrices are of the same dimensions as N and P, respectively.

There are other equivalent definitions of the matrix sign function, but the one given here is especially useful in deriving many of its key properties. The JCF definition of the matrix sign function does not generally lend itself to reliable computation on a finite-word-length digital computer. In fact, its reliable numerical calculation is an interesting topic in its own right.

We state some of the more useful properties of the matrix sign function as theorems. Their straightforward proofs are left to the exercises.

Theorem 9.42. *Suppose $A \in \mathbb{C}^{n \times n}$ has no eigenvalues on the imaginary axis, and let $S = \mathrm{sgn}(A)$. Then the following hold:*

1. *S is diagonalizable with eigenvalues equal to ± 1.*
2. *$S^2 = I$.*
3. *$AS = SA$.*
4. *$\mathrm{sgn}(A^H) = (\mathrm{sgn}(A))^H$.*
5. *$\mathrm{sgn}(T^{-1}AT) = T^{-1}\mathrm{sgn}(A)T$ for all nonsingular $T \in \mathbb{C}^{n \times n}$.*
6. *$\mathrm{sgn}(cA) = \mathrm{sgn}(c)\,\mathrm{sgn}(A)$ for all nonzero real scalars c.*

Theorem 9.43. *Suppose $A \in \mathbb{C}^{n \times n}$ has no eigenvalues on the imaginary axis, and let $S = \mathrm{sgn}(A)$. Then the following hold:*

1. *$\mathcal{R}(S - I)$ is an A-invariant subspace corresponding to the left half-plane eigenvalues of A (the* **negative invariant subspace***).*
2. *$\mathcal{R}(S + I)$ is an A-invariant subspace corresponding to the right half-plane eigenvalues of A (the* **positive invariant subspace***).*
3. *$\mathrm{neg}A \equiv (I - S)/2$ is a projection onto the negative invariant subspace of A.*
4. *$\mathrm{pos}A \equiv (I + S)/2$ is a projection onto the positive invariant subspace of A.*

EXERCISES

1. Let $A \in \mathbb{C}^{n \times n}$ have distinct eigenvalues $\lambda_1, \ldots, \lambda_n$ with corresponding right eigenvectors $x_1, \ldots, x_n$ and left eigenvectors $y_1, \ldots, y_n$, respectively. Let $v \in \mathbb{C}^n$ be an arbitrary vector. Show that v can be expressed (uniquely) as a linear combination

$$v = \sum_{i=1}^{n} \frac{y_i^H v}{y_i^H x_i} x_i$$

of the right eigenvectors. Find the appropriate expression for v as a linear combination of the left eigenvectors as well.

2. Suppose $A \in \mathbb{C}^{n\times n}$ is skew-Hermitian, i.e., $A^H = -A$. Prove that all eigenvalues of a skew-Hermitian matrix must be pure imaginary.

3. Suppose $A \in \mathbb{C}^{n\times n}$ is Hermitian. Let λ be an eigenvalue of A with corresponding right eigenvector x. Show that x is also a left eigenvector for λ. Prove the same result if A is skew-Hermitian.

4. Suppose a matrix $A \in \mathbb{R}^{5\times 5}$ has eigenvalues $\{2, 2, 2, 2, 3\}$. Determine all possible JCFs for A.

5. Determine the eigenvalues, right eigenvectors and right principal vectors if necessary, and (real) JCFs of the following matrices:

$$\text{(a)} \begin{bmatrix} 2 & -1 \\ 1 & 0 \end{bmatrix}, \quad \text{(b)} \begin{bmatrix} 2 & 1 \\ 1 & 2 \end{bmatrix}, \quad \text{(c)} \begin{bmatrix} 4 & 5 \\ -1 & 0 \end{bmatrix}.$$

6. Determine the JCFs of the following matrices:

$$\text{(a)} \begin{bmatrix} 0 & 1 & 0 \\ 0 & 0 & 1 \\ 1 & -3 & 3 \end{bmatrix}, \quad \text{(b)} \begin{bmatrix} 2 & -2 & -1 \\ 1 & -1 & -1 \\ -1 & 2 & 2 \end{bmatrix}.$$

7. Let

$$A = \begin{bmatrix} 1 & \frac{1}{2} & \frac{1}{2} \\ 0 & \frac{1}{2} & -\frac{1}{2} \\ 0 & \frac{1}{2} & \frac{3}{2} \end{bmatrix}.$$

Find a nonsingular matrix X such that $X^{-1}AX = J$, where J is the JCF

$$J = \begin{bmatrix} 1 & 0 & 0 \\ 0 & 1 & 1 \\ 0 & 0 & 1 \end{bmatrix}.$$

Hint: Use $[-1 \;\; 1 \;\; -1]^T$ as an eigenvector. The vectors $[0 \;\; 1 \;\; -1]^T$ and $[1 \;\; 0 \;\; 0]^T$ are both eigenvectors, but then the equation $(A - I)x^{(2)} = x^{(1)}$ can't be solved.

8. Show that all right eigenvectors of the Jordan block matrix in Theorem 9.30 must be multiples of $e_1 \in \mathbb{R}^k$. Characterize all left eigenvectors.

9. Let $A \in \mathbb{R}^{n\times n}$ be of the form $A = xy^T$, where $x, y \in \mathbb{R}^n$ are nonzero vectors with $x^T y = 0$. Determine the JCF of A.

10. Let $A \in \mathbb{R}^{n\times n}$ be of the form $A = I + xy^T$, where $x, y \in \mathbb{R}^n$ are nonzero vectors with $x^T y = 0$. Determine the JCF of A.

11. Suppose a matrix $A \in \mathbb{R}^{16\times 16}$ has 16 eigenvalues at 0 and its JCF consists of a single Jordan block of the form specified in Theorem 9.22. Suppose the small number 10^{-16} is added to the (16,1) element of J. What are the eigenvalues of this slightly perturbed matrix?

12. Show that every matrix $A \in \mathbb{R}^{n\times n}$ can be factored in the form $A = S_1 S_2$, where S_1 and S_2 are real symmetric matrices and one of them, say S_1, is nonsingular.
Hint: Suppose $A = XJX^{-1}$ is a reduction of A to JCF and suppose we can construct the "symmetric factorization" of J. Then $A = (XS_1X^T)(X^{-T}S_2X^{-1})$ would be the required symmetric factorization of A. Thus, it suffices to prove the result for the JCF. The transformation P in (9.18) is useful.

13. Prove that every matrix $A \in \mathbb{R}^{n\times n}$ is similar to its transpose and determine a similarity transformation explicitly.
Hint: Use the factorization in the previous exercise.

14. Consider the block upper triangular matrix

$$A = \begin{bmatrix} A_{11} & A_{12} \\ 0 & A_{22} \end{bmatrix},$$

where $A \in \mathbb{R}^{n\times n}$ and $A_{11} \in \mathbb{R}^{k\times k}$ with $1 \le k \le n$. Suppose $A_{12} \neq 0$ and that we want to block diagonalize A via the similarity transformation

$$T = \begin{bmatrix} I & X \\ 0 & I \end{bmatrix},$$

where $X \in \mathbb{R}^{k\times(n-k)}$, i.e.,

$$T^{-1}AT = \begin{bmatrix} A_{11} & 0 \\ 0 & A_{22} \end{bmatrix}.$$

Find a matrix equation that X must satisfy for this to be possible. If $n = 2$ and $k = 1$, what can you say further, in terms of A_{11} and A_{22}, about when the equation for X is solvable?

15. Prove Theorem 9.42.

16. Prove Theorem 9.43.

17. Suppose $A \in \mathbb{C}^{n\times n}$ has all its eigenvalues in the left half-plane. Prove that $\operatorname{sgn}(A) = -I$.

Chapter 10

Canonical Forms

10.1 Some Basic Canonical Forms

Problem: Let $\mathcal{V}$ and $\mathcal{W}$ be vector spaces and suppose $A : \mathcal{V} \to \mathcal{W}$ is a linear transformation. Find bases in $\mathcal{V}$ and $\mathcal{W}$ with respect to which Mat A has a "simple form" or "canonical form." In matrix terms, if $A \in \mathbb{R}^{m\times n}$, find $P \in \mathbb{R}_m^{m\times m}$ and $Q \in \mathbb{R}_n^{n\times n}$ such that PAQ has a "canonical form." The transformation $A \mapsto PAQ$ is called an **equivalence**; it is called an **orthogonal equivalence** if P and Q are orthogonal matrices.

Remark 10.1. We can also consider the case $A \in \mathbb{C}^{m\times n}$ and **unitary equivalence** if P and Q are unitary.

Two special cases are of interest:

1. If $\mathcal{W} = \mathcal{V}$ and $Q = P^{-1}$, the transformation $A \mapsto PAP^{-1}$ is called a **similarity**.
2. If $\mathcal{W} = \mathcal{V}$ and if $Q = P^T$ is orthogonal, the transformation $A \mapsto PAP^T$ is called an **orthogonal similarity** (or **unitary similarity** in the complex case).

The following results are typical of what can be achieved under a unitary similarity. If $A = A^H \in \mathbb{C}^{n\times n}$ has eigenvalues $\lambda_1, \ldots, \lambda_n$, then there exists a unitary matrix U such that $U^H AU = D$, where $D = \operatorname{diag}(\lambda_1, \ldots, \lambda_n)$. This is proved in Theorem 10.2. What other matrices are "diagonalizable" under unitary similarity? The answer is given in Theorem 10.9, where it is proved that a general matrix $A \in \mathbb{C}^{n\times n}$ is unitarily similar to a diagonal matrix if and only if it is normal (i.e., $AA^H = A^H A$). Normal matrices include Hermitian, skew-Hermitian, and unitary matrices (and their "real" counterparts: symmetric, skew-symmetric, and orthogonal, respectively), as well as other matrices that merely satisfy the definition, such as $A = \begin{bmatrix} a & b \\ -b & a \end{bmatrix}$ for real scalars a and b. If a matrix A is not normal, the most "diagonal" we can get is the JCF described in Chapter 9.

Theorem 10.2. *Let $A = A^H \in \mathbb{C}^{n\times n}$ have (real) eigenvalues $\lambda_1, \ldots, \lambda_n$. Then there exists a unitary matrix X such that $X^H AX = D = \operatorname{diag}(\lambda_1, \ldots, \lambda_n)$ (the columns of X are orthonormal eigenvectors for A).*

Proof: Let x_1 be a right eigenvector corresponding to λ_1, and normalize it such that $x_1^H x_1 = 1$. Then there exist $n-1$ additional vectors $x_2, \ldots, x_n$ such that $X = [x_1, \ldots, x_n] = [x_1 \;\; X_2]$ is unitary. Now

$$X^H A X = \begin{bmatrix} x_1^H \\ X_2^H \end{bmatrix} A \, [x_1 \;\; X_2] = \begin{bmatrix} x_1^H A x_1 & x_1^H A X_2 \\ X_2^H A x_1 & X_2^H A X_2 \end{bmatrix}$$

$$= \begin{bmatrix} \lambda_1 & x_1^H A X_2 \\ 0 & X_2^H A X_2 \end{bmatrix} \tag{10.1}$$

$$= \begin{bmatrix} \lambda_1 & 0 \\ 0 & X_2^H A X_2 \end{bmatrix}. \tag{10.2}$$

In (10.1) we have used the fact that $Ax_1 = \lambda_1 x_1$. When combined with the fact that $x_1^H x_1 = 1$, we get λ_1 remaining in the (1,1)-block. We also get 0 in the (2,1)-block by noting that x_1 is orthogonal to all vectors in X_2. In (10.2), we get 0 in the (1,2)-block by noting that $X^H A X$ is Hermitian. The proof is completed easily by induction upon noting that the (2,2)-block must have eigenvalues $\lambda_2, \ldots, \lambda_n$. □

Given a unit vector $x_1 \in \mathbb{R}^n$, the construction of $X_2 \in \mathbb{R}^{n \times (n-1)}$ such that $X = [x_1 \;\; X_2]$ is orthogonal is frequently required. The construction can actually be performed quite easily by means of Householder (or Givens) transformations as in the proof of the following general result.

Theorem 10.3. *Let $X_1 \in \mathbb{C}^{n \times k}$ have orthonormal columns and suppose U is a unitary matrix such that $UX_1 = \left[\begin{smallmatrix} R \\ 0 \end{smallmatrix}\right]$, where $R \in \mathbb{C}^{k \times k}$ is upper triangular. Write $U^H = [U_1 \;\; U_2]$ with $U_1 \in \mathbb{C}^{n \times k}$. Then $[X_1 \;\; U_2]$ is unitary.*

Proof: Let $X_1 = [x_1, \ldots, x_k]$. Construct a sequence of Householder matrices (also known as elementary reflectors) $H_1, \ldots, H_k$ in the usual way (see below) such that

$$H_k \cdots H_1 [x_1, \ldots, x_k] = \begin{bmatrix} R \\ 0 \end{bmatrix},$$

where R is upper triangular (and nonsingular since $x_1, \ldots, x_k$ are orthonormal). Let $U = H_k \cdots H_1$. Then $U^H = H_1 \cdots H_k$ and

$$X^H U^H = \begin{bmatrix} x_1^H \\ \vdots \\ x_k^H \end{bmatrix} [U_1 \;\; U_2] = \begin{bmatrix} R^H & 0 \end{bmatrix}.$$

Then $x_i^H U_2 = 0$ ($i \in \underline{k}$) means that x_i is orthogonal to each of the $n-k$ columns of U_2. But the latter are orthonormal since they are the last $n-k$ rows of the unitary matrix U. Thus, $[X_1 \;\; U_2]$ is unitary. □

The construction called for in Theorem 10.2 is then a special case of Theorem 10.3 for $k = 1$. We illustrate the construction of the necessary Householder matrix for $k = 1$. For simplicity, we consider the real case. Let the unit vector x_1 be denoted by $[\xi_1, \ldots, \xi_n]^T$.

Then the necessary Householder matrix needed for the construction of X_2 is given by $U = I - 2uu^+ = I - \frac{2}{u^T u} uu^T$, where $u = [\xi_1 \pm 1, \xi_2, \ldots, \xi_n]^T$. It can easily be checked that U is symmetric and $U^T U = U^2 = I$, so U is orthogonal. To see that U effects the necessary compression of x_1, it is easily verified that $u^T u = 2 \pm 2\xi_1$ and $u^T x_1 = 1 \pm \xi_1$. Thus,

$$\begin{aligned} Ux_1 &= \left(I - \frac{2}{u^T u} uu^T \right) x_1 \\ &= x_1 - \frac{2u^T x_1}{u^T u} u \\ &= \begin{bmatrix} \xi_1 \\ \xi_2 \\ \vdots \\ \xi_n \end{bmatrix} - 1 \cdot \begin{bmatrix} \xi_1 \pm 1 \\ \xi_2 \\ \vdots \\ \xi_n \end{bmatrix} \\ &= \begin{bmatrix} \mp 1 \\ 0 \\ \vdots \\ 0 \end{bmatrix}. \end{aligned}$$

Further details on Householder matrices, including the choice of sign and the complex case, can be consulted in standard numerical linear algebra texts such as [7], [11], [23], [25].

The real version of Theorem 10.2 is worth stating separately since it is applied frequently in applications.

Theorem 10.4. *Let $A = A^T \in \mathbb{R}^{n \times n}$ have eigenvalues $\lambda_1, \ldots, \lambda_n$. Then there exists an orthogonal matrix $X \in \mathbb{R}^{n \times n}$ (whose columns are orthonormal eigenvectors of A) such that $X^T A X = D = \operatorname{diag}(\lambda_1, \ldots, \lambda_n)$.*

Note that Theorem 10.4 implies that a symmetric matrix A (with the obvious analogue from Theorem 10.2 for Hermitian matrices) can be written

$$A = XDX^T = \sum_{i=1}^{n} \lambda_i x_i x_i^T, \tag{10.3}$$

which is often called the **spectral representation** of A. In fact, A in (10.3) is actually a weighted sum of orthogonal projections P_i (onto the one-dimensional eigenspaces corresponding to the λ_i's), i.e.,

$$A = \sum_{i=1}^{n} \lambda_i P_i,$$

where $P_i = P_{\mathcal{R}(x_i)} = x_i x_i^+ = x_i x_i^T$ since $x_i^T x_i = 1$.

The following pair of theorems form the theoretical foundation of the double-Francis-QR algorithm used to compute matrix eigenvalues in a numerically stable and reliable way.

Theorem 10.5 (Schur). *Let $A \in \mathbb{C}^{n \times n}$. Then there exists a unitary matrix U such that $U^H AU = T$, where T is upper triangular.*

Proof: The proof of this theorem is essentially the same as that of Theorem 10.2 except that in this case (using the notation U rather than X) the (1,2)-block $u_1^H AU_2$ is not 0. □

In the case of $A \in \mathbb{R}^{n \times n}$, it is thus unitarily similar to an upper triangular matrix, but if A has a complex conjugate pair of eigenvalues, then complex arithmetic is clearly needed to place such eigenvalues on the diagonal of T. However, the next theorem shows that every $A \in \mathbb{R}^{n \times n}$ is also orthogonally similar (i.e., real arithmetic) to a **quasi-upper-triangular matrix**. A quasi-upper-triangular matrix is block upper triangular with 1×1 diagonal blocks corresponding to its real eigenvalues and 2×2 diagonal blocks corresponding to its complex conjugate pairs of eigenvalues.

Theorem 10.6 (Murnaghan–Wintner). *Let $A \in \mathbb{R}^{n \times n}$. Then there exists an orthogonal matrix U such that $U^T AU = S$, where S is quasi-upper-triangular.*

Definition 10.7. *The triangular matrix T in Theorem 10.5 is called a* **Schur canonical form** *or Schur form. The quasi-upper-triangular matrix S in Theorem 10.6 is called a* **real Schur canonical form** *or real Schur form (RSF). The columns of a unitary [orthogonal] matrix U that reduces a matrix to [real] Schur form are called* **Schur vectors**.

Example 10.8. The matrix

$$S = \begin{bmatrix} -2 & 5 & 8 \\ -2 & 4 & 4 \\ 0 & 0 & 2 \end{bmatrix}$$

is in RSF. Its real JCF is

$$J = \begin{bmatrix} 1 & 1 & 0 \\ -1 & 1 & 0 \\ 0 & 0 & 2 \end{bmatrix}.$$

Note that only the first Schur vector (and then only if the corresponding first eigenvalue is real if U is orthogonal) is an eigenvector. However, what is true, and sufficient for virtually all applications (see, for example, [17]), is that the first k Schur vectors span the same A-invariant subspace as the eigenvectors corresponding to the first k eigenvalues along the diagonal of T (or S).

While every matrix can be reduced to Schur form (or RSF), it is of interest to know when we can go further and reduce a matrix via unitary similarity to diagonal form. The following theorem answers this question.

Theorem 10.9. *A matrix $A \in \mathbb{C}^{n \times n}$ is unitarily similar to a diagonal matrix if and only if A is normal (i.e., $A^H A = AA^H$).*

Proof: Suppose U is a unitary matrix such that $U^H AU = D$, where D is diagonal. Then

$$AA^H = UDU^H UD^H U^H = UDD^H U^H = UD^H DU^H = A^H A$$

so A is normal.

Conversely, suppose A is normal and let U be a unitary matrix such that $U^H AU = T$, where T is an upper triangular matrix (Theorem 10.5). Then

$$TT^H = U^H AUU^H A^H U = U^H AA^H U = U^H A^H AU = T^H T.$$

It is then a routine exercise to show that T must, in fact, be diagonal. □

10.2 Definite Matrices

Definition 10.10. *A symmetric matrix $A \in \mathbb{R}^{n\times n}$ is*

1. **positive definite** *if and only if $x^T Ax > 0$ for all nonzero $x \in \mathbb{R}^n$. We write $A > 0$.*
2. **nonnegative definite** (*or* **positive semidefinite**) *if and only if $x^T Ax \geq 0$ for all nonzero $x \in \mathbb{R}^n$. We write $A \geq 0$.*
3. **negative definite** *if $-A$ is positive definite. We write $A < 0$.*
4. **nonpositive definite** (*or* **negative semidefinite**) *if $-A$ is nonnegative definite. We write $A \leq 0$.*

Also, if A and B are symmetric matrices, we write $A > B$ if and only if $A - B > 0$ or $B - A < 0$. Similarly, we write $A \geq B$ if and only if $A - B \geq 0$ or $B - A \leq 0$.

Remark 10.11. If $A \in \mathbb{C}^{n\times n}$ is Hermitian, all the above definitions hold except that superscript Hs replace Ts. Indeed, this is generally true for all results in the remainder of this section that may be stated in the real case for simplicity.

Remark 10.12. If a matrix is neither definite nor semidefinite, it is said to be **indefinite**.

Theorem 10.13. *Let $A = A^H \in \mathbb{C}^{n\times n}$ with eigenvalues $\lambda_1 \geq \lambda_2 \geq \cdots \geq \lambda_n$. Then for all $x \in \mathbb{C}^n$,*

$$\lambda_n x^H x \leq x^H Ax \leq \lambda_1 x^H x.$$

Proof: Let U be a unitary matrix that diagonalizes A as in Theorem 10.2. Furthermore, let $y = U^H x$, where x is an arbitrary vector in $\mathbb{C}^n$, and denote the components of y by η_i, $i \in \underline{n}$. Then

$$x^H Ax = (U^H x)^H U^H AU(U^H x) = y^H Dy = \sum_{i=1}^{n} \lambda_i |\eta_i|^2.$$

But clearly

$$\sum_{i=1}^{n} \lambda_i |\eta_i|^2 \leq \lambda_1 y^H y = \lambda_1 x^H x$$

and

$$\sum_{i=1}^{n} \lambda_i |\eta_i|^2 \geq \lambda_n y^H y = \lambda_n x^H x,$$

from which the theorem follows. □

Remark 10.14. The ratio $\frac{x^H A x}{x^H x}$ for $A = A^H \in \mathbb{C}^{n \times n}$ and nonzero $x \in \mathbb{C}^n$ is called the **Rayleigh quotient** of x. Theorem 10.13 provides upper (λ_1) and lower (λ_n) bounds for the Rayleigh quotient. If $A = A^H \in \mathbb{C}^{n \times n}$ is positive definite, $x^H A x > 0$ for all nonzero $x \in \mathbb{C}^n$, so $0 < \lambda_n \leq \cdots \leq \lambda_1$.

Corollary 10.15. *Let $A \in \mathbb{C}^{n \times n}$. Then $\|A\|_2 = \lambda_{\max}^{\frac{1}{2}}(A^H A)$.*

Proof: For all $x \in \mathbb{C}^n$ we have

$$\frac{\|Ax\|_2}{\|x\|_2} = \left(\frac{x^H A^H A x}{x^H x} \right)^{\frac{1}{2}} \leq \lambda_{\max}^{\frac{1}{2}}(A^H A).$$

Let x be an eigenvector corresponding to $\lambda_{\max}(A^H A)$. Then $\frac{\|Ax\|_2}{\|x\|_2} = \lambda_{\max}^{\frac{1}{2}}(A^H A)$, whence

$$\|A\|_2 = \max_{x \neq 0} \frac{\|Ax\|_2}{\|x\|_2} = \lambda_{\max}^{\frac{1}{2}}(A^H A). \quad \square$$

Definition 10.16. *A* **principal submatrix** *of an $n \times n$ matrix A is the $(n-k) \times (n-k)$ matrix that remains by deleting k rows and the corresponding k columns. A* **leading principal submatrix** *of order $n - k$ is obtained by deleting the last k rows and columns.*

Theorem 10.17. *A symmetric matrix $A \in \mathbb{R}^{n \times n}$ is positive definite if and only if any of the following three equivalent conditions hold:*

1. *The determinants of all leading principal submatrices of A are positive.*

2. *All eigenvalues of A are positive.*

3. *A can be written in the form $M^T M$, where $M \in \mathbb{R}^{n \times n}$ is nonsingular.*

Theorem 10.18. *A symmetric matrix $A \in \mathbb{R}^{n \times n}$ is nonnegative definite if and only if any of the following three equivalent conditions hold:*

1. *The determinants of all principal submatrices of A are nonnegative.*

2. *All eigenvalues of A are nonnegative.*

3. *A can be written in the form $M^T M$, where $M \in \mathbb{R}^{k \times n}$ and $k \geq \operatorname{rank}(A) = \operatorname{rank}(M)$.*

Remark 10.19. Note that the determinants of *all* principal submatrices must be nonnegative in Theorem 10.18.1, not just those of the leading principal submatrices. For example, consider the matrix $A = \begin{bmatrix} 0 & 0 \\ 0 & -1 \end{bmatrix}$. The determinant of the 1×1 leading submatrix is 0 and the determinant of the 2×2 leading submatrix is also 0 (cf. Theorem 10.17). However, the principal submatrix consisting of the (2,2) element is, in fact, negative and A is nonpositive definite.

Remark 10.20. The factor M in Theorem 10.18.3 is not unique. For example, if

$$A = \begin{bmatrix} 1 & 0 \\ 0 & 0 \end{bmatrix},$$

then M can be

$$[1 \quad 0], \begin{bmatrix} \frac{1}{\sqrt{2}} & 0 \\ \frac{1}{\sqrt{2}} & 0 \end{bmatrix}, \begin{bmatrix} \frac{1}{\sqrt{3}} & 0 \\ \frac{1}{\sqrt{3}} & 0 \\ \frac{1}{\sqrt{3}} & 0 \end{bmatrix}, \ldots .$$

Recall that $A \geq B$ if the matrix $A - B$ is nonnegative definite. The following theorem is useful in "comparing" symmetric matrices. Its proof is straightforward from basic definitions.

Theorem 10.21. *Let $A, B \in \mathbb{R}^{n \times n}$ be symmetric.*

1. *If $A \geq B$ and $M \in \mathbb{R}^{n \times m}$, then $M^T AM \geq M^T BM$.*
2. *If $A > B$ and $M \in \mathbb{R}_m^{n \times m}$, then $M^T AM > M^T BM$.*

The following standard theorem is stated without proof (see, for example, [16, p. 181]). It concerns the notion of the "square root" of a matrix. That is, if $A \in \mathbb{R}^{n \times n}$, we say that $S \in \mathbb{R}^{n \times n}$ is a **square root** of A if $S^2 = A$. In general, matrices (both symmetric and nonsymmetric) have infinitely many square roots. For example, if $A = I_2$, any matrix S of the form $\left[\begin{smallmatrix} \cos\theta & \sin\theta \\ \sin\theta & -\cos\theta \end{smallmatrix}\right]$ is a square root.

Theorem 10.22. *Let $A \in \mathbb{R}^{n \times n}$ be nonnegative definite. Then A has a unique nonnegative definite square root S. Moreover, $SA = AS$ and* $\operatorname{rank} S = \operatorname{rank} A$ *(and hence S is positive definite if A is positive definite).*

A stronger form of the third characterization in Theorem 10.17 is available and is known as the **Cholesky factorization**. It is stated and proved below for the more general Hermitian case.

Theorem 10.23. *Let $A \in \mathbb{C}^{n \times n}$ be Hermitian and positive definite. Then there exists a unique nonsingular lower triangular matrix L with positive diagonal elements such that $A = LL^H$.*

Proof: The proof is by induction. The case $n = 1$ is trivially true. Write the matrix A in the form

$$A = \begin{bmatrix} B & b \\ b^H & a_{nn} \end{bmatrix}.$$

By our induction hypothesis, assume the result is true for matrices of order $n - 1$ so that B may be written as $B = L_1 L_1^H$, where $L_1 \in \mathbb{C}^{(n-1)\times(n-1)}$ is nonsingular and lower triangular with positive diagonal elements. It remains to prove that we can write the $n \times n$ matrix A in the form

$$A = \begin{bmatrix} L_1 L_1^H & b \\ b^H & a_{nn} \end{bmatrix} = \begin{bmatrix} L_1 & 0 \\ c^H & \alpha \end{bmatrix} \begin{bmatrix} L_1^H & c \\ 0 & \alpha \end{bmatrix},$$

where α is positive. Performing the indicated matrix multiplication and equating the corresponding submatrices, we see that we must have $L_1 c = b$ and $a_{nn} = c^H c + \alpha^2$. Clearly c is given simply by $c = L_1^{-1} b$. Substituting in the expression involving α, we find $\alpha^2 = a_{nn} - b^H L_1^{-H} L_1^{-1} b = a_{nn} - b^H B^{-1} b$ (= the Schur complement of B in A). But we know that

$$0 < \det(A) = \det \begin{bmatrix} B & b \\ b^H & a_{nn} \end{bmatrix} = \det(B) \det(a_{nn} - b^H B^{-1} b).$$

Since $\det(B) > 0$, we must have $a_{nn} - b^H B^{-1} b > 0$. Choosing α to be the positive square root of $a_{nn} - b^H B^{-1} b$ completes the proof. $\square$

10.3 Equivalence Transformations and Congruence

Theorem 10.24. *Let $A \in \mathbb{C}_r^{m \times n}$. Then there exist matrices $P \in \mathbb{C}_m^{m \times m}$ and $Q \in \mathbb{C}_n^{n \times n}$ such that*

$$PAQ = \begin{bmatrix} I_r & 0 \\ 0 & 0 \end{bmatrix}. \tag{10.4}$$

Proof: A classical proof can be consulted in, for example, [21, p. 131]. Alternatively, suppose A has an SVD of the form (5.2) in its complex version. Then

$$\begin{bmatrix} S^{-1} & 0 \\ 0 & I \end{bmatrix} \begin{bmatrix} U_1^H \\ U_2^H \end{bmatrix} AV = \begin{bmatrix} I & 0 \\ 0 & 0 \end{bmatrix}.$$

Take $P = \begin{bmatrix} S^{-1} U_1^H \\ U_2^H \end{bmatrix}$ and $Q = V$ to complete the proof. $\square$

Note that the greater freedom afforded by the equivalence transformation of Theorem 10.24, as opposed to the more restrictive situation of a similarity transformation, yields a far "simpler" canonical form (10.4). However, numerical procedures for computing such an equivalence directly via, say, Gaussian or elementary row and column operations, are generally unreliable. The numerically preferred equivalence is, of course, the unitary equivalence known as the SVD. However, the SVD is relatively expensive to compute and other canonical forms exist that are intermediate between (10.4) and the SVD; see, for example [7, Ch. 5], [4, Ch. 2]. Two such forms are stated here. They are more stably computable than (10.4) and more efficiently computable than a full SVD. Many similar results are also available.

Theorem 10.25 (Complete Orthogonal Decomposition). *Let $A \in \mathbb{C}_r^{m\times n}$. Then there exist unitary matrices $U \in \mathbb{C}^{m\times m}$ and $V \in \mathbb{C}^{n\times n}$ such that*

$$UAV = \begin{bmatrix} R & 0 \\ 0 & 0 \end{bmatrix}, \tag{10.5}$$

where $R \in \mathbb{C}_r^{r\times r}$ is upper (or lower) triangular with positive diagonal elements.

Proof: For the proof, see [4]. □

Theorem 10.26. *Let $A \in \mathbb{C}_r^{m\times n}$. Then there exists a unitary matrix $Q \in \mathbb{C}^{m\times m}$ and a permutation matrix $\Pi \in \mathbb{C}^{n\times n}$ such that*

$$QA\Pi = \begin{bmatrix} R & S \\ 0 & 0 \end{bmatrix}, \tag{10.6}$$

where $R \in \mathbb{C}_r^{r\times r}$ is upper triangular and $S \in \mathbb{C}^{r\times(n-r)}$ is arbitrary but in general nonzero.

Proof: For the proof, see [4]. □

Remark 10.27. When A has full column rank but is "near" a rank deficient matrix, various **rank revealing QR decompositions** are available that can sometimes detect such phenomena at a cost considerably less than a full SVD. Again, see [4] for details.

Definition 10.28. *Let $A \in \mathbb{C}^{n\times n}$ and $X \in \mathbb{C}_n^{n\times n}$. The transformation $A \mapsto X^H AX$ is called a* **congruence**. *Note that a congruence is a similarity if and only if X is unitary.*

Note that congruence preserves the property of being Hermitian; i.e., if A is Hermitian, then $X^H AX$ is also Hermitian. It is of interest to ask what other properties of a matrix are preserved under congruence. It turns out that the principal property so preserved is the sign of each eigenvalue.

Definition 10.29. *Let $A = A^H \in \mathbb{C}^{n\times n}$ and let π, ν, and ζ denote the numbers of positive, negative, and zero eigenvalues, respectively, of A. Then the* **inertia** *of A is the triple of numbers* $\mathrm{In}(A) = (\pi, \nu, \zeta)$. *The* **signature** *of A is given by* $\mathrm{sig}(A) = \pi - \nu$.

Example 10.30.

1. $\mathrm{In}\begin{bmatrix} 2 & 1 & 0 & 0 \\ 1 & 1 & 0 & 0 \\ 0 & 0 & -1 & 0 \\ 0 & 0 & 0 & 0 \end{bmatrix} = (2, 1, 1).$

2. If $A = A^H \in \mathbb{C}^{n\times n}$, then $A > 0$ if and only if $\mathrm{In}(A) = (n, 0, 0)$.

3. If $\mathrm{In}(A) = (\pi, \nu, \zeta)$, then $\mathrm{rank}(A) = \pi + \nu$.

Theorem 10.31 (Sylvester's Law of Inertia). *Let $A = A^H \in \mathbb{C}^{n\times n}$ and $X \in \mathbb{C}_n^{n\times n}$. Then* $\mathrm{In}(A) = \mathrm{In}(X^H AX)$.

Proof: For the proof, see, for example, [21, p. 134]. □

Theorem 10.31 guarantees that rank and signature of a matrix are preserved under congruence. We then have the following.

Theorem 10.32. *Let $A = A^H \in \mathbb{C}^{n\times n}$ with $\mathrm{In}(A) = (\pi, \nu, \zeta)$. Then there exists a matrix $X \in \mathbb{C}_n^{n\times n}$ such that $X^H AX = \mathrm{diag}(1, \dots, 1, -1, \dots, -1, 0, \dots, 0)$, where the number of 1's is π, the number of -1's is ν, and the number of 0's is ζ.*

Proof: Let $\lambda_1, \dots, \lambda_n$ denote the eigenvalues of A and order them such that the first π are positive, the next ν are negative, and the final ζ are 0. By Theorem 10.2 there exists a unitary matrix U such that $U^H AU = \mathrm{diag}(\lambda_1, \dots, \lambda_n)$. Define the $n \times n$ matrix

$$W = \mathrm{diag}(1/\sqrt{\lambda_1}, \dots, 1/\sqrt{\lambda_\pi}, 1/\sqrt{-\lambda_{\pi+1}}, \dots, 1/\sqrt{-\lambda_{\pi+\nu}}, 1, \dots, 1).$$

Then it is easy to check that $X = UW$ yields the desired result. □

10.3.1 Block matrices and definiteness

Theorem 10.33. *Suppose $A = A^T$ and $D = D^T$. Then*

$$\begin{bmatrix} A & B \\ B^T & D \end{bmatrix} > 0$$

if and only if either $A > 0$ and $D - B^T A^{-1} B > 0$, or $D > 0$ and $A - BD^{-1}B^T > 0$.

Proof: The proof follows by considering, for example, the congruence

$$\begin{bmatrix} A & B \\ B^T & D \end{bmatrix} \mapsto \begin{bmatrix} I & -A^{-1}B \\ 0 & I \end{bmatrix}^T \begin{bmatrix} A & B \\ B^T & D \end{bmatrix} \begin{bmatrix} I & -A^{-1}B \\ 0 & I \end{bmatrix}.$$

The details are straightforward and are left to the reader. □

Remark 10.34. Note the symmetric Schur complements of A (or D) in the theorem.

Theorem 10.35. *Suppose $A = A^T$ and $D = D^T$. Then*

$$\begin{bmatrix} A & B \\ B^T & D \end{bmatrix} \geq 0$$

if and only if $A \geq 0$, $AA^+B = B$, and $D - B^T A^+ B \geq 0$.

Proof: Consider the congruence with

$$\begin{bmatrix} I & -A^+B \\ 0 & I \end{bmatrix}$$

and proceed as in the proof of Theorem 10.33. □

10.4 Rational Canonical Form

One final canonical form to be mentioned is the **rational canonical form**.

Definition 10.36. *A matrix $A \in \mathbb{R}^{n\times n}$ is said to be* **nonderogatory** *if its minimal polynomial and characteristic polynomial are the same or, equivalently, if its Jordan canonical form has only one block associated with each distinct eigenvalue.*

Suppose $A \in \mathbb{R}^{n\times n}$ is a nonderogatory matrix and suppose its characteristic polynomial is $\pi(\lambda) = \lambda^n - (a_0 + a_1\lambda + \cdots + a_{n-1}\lambda^{n-1})$. Then it can be shown (see [12]) that A is similar to a matrix of the form

$$\begin{bmatrix} 0 & 1 & 0 & \cdots & 0 \\ 0 & 0 & 1 & \ddots & \vdots \\ \vdots & & \ddots & \ddots & 0 \\ 0 & \cdots & \cdots & 0 & 1 \\ a_0 & a_1 & \cdots & & a_{n-1} \end{bmatrix}. \tag{10.7}$$

Definition 10.37. *A matrix $A \in \mathbb{R}^{n\times n}$ of the form (10.7) is called a* **companion matrix** *or is said to be in* **companion form**.

Companion matrices also appear in the literature in several equivalent forms. To illustrate, consider the companion matrix

$$A = \begin{bmatrix} 0 & 1 & 0 & 0 \\ 0 & 0 & 1 & 0 \\ 0 & 0 & 0 & 1 \\ a_0 & a_1 & a_2 & a_3 \end{bmatrix}. \tag{10.8}$$

This matrix is a special case of a matrix in lower Hessenberg form. Using the reverse-order identity similarity P given by (9.18), A is easily seen to be similar to the following matrix in upper Hessenberg form:

$$\begin{bmatrix} a_3 & a_2 & a_1 & a_0 \\ 1 & 0 & 0 & 0 \\ 0 & 1 & 0 & 0 \\ 0 & 0 & 1 & 0 \end{bmatrix}. \tag{10.9}$$

Moreover, since a matrix is similar to its transpose (see exercise 13 in Chapter 9), the following are also companion matrices similar to the above:

$$\begin{bmatrix} 0 & 0 & 0 & a_0 \\ 1 & 0 & 0 & a_1 \\ 0 & 1 & 0 & a_2 \\ 0 & 0 & 1 & a_3 \end{bmatrix}, \quad \begin{bmatrix} a_3 & 1 & 0 & 0 \\ a_2 & 0 & 1 & 0 \\ a_1 & 0 & 0 & 1 \\ a_0 & 0 & 0 & 0 \end{bmatrix}. \tag{10.10}$$

Notice that in all cases a companion matrix is nonsingular if and only if $a_0 \neq 0$. In fact, the inverse of a nonsingular companion matrix is again in companion form. For example,

$$\begin{bmatrix} 0 & 1 & 0 & 0 \\ 0 & 0 & 1 & 0 \\ 0 & 0 & 0 & 1 \\ a_0 & a_1 & a_2 & a_3 \end{bmatrix}^{-1} = \begin{bmatrix} -\frac{a_1}{a_0} & -\frac{a_2}{a_0} & -\frac{a_3}{a_0} & \frac{1}{a_0} \\ 1 & 0 & 0 & 0 \\ 0 & 1 & 0 & 0 \\ 0 & 0 & 1 & 0 \end{bmatrix} \tag{10.11}$$

with a similar result for companion matrices of the form (10.10).

If a companion matrix of the form (10.7) is singular, i.e., if $a_0 = 0$, then its pseudoinverse can still be computed. Let $a \in \mathbb{R}^{n-1}$ denote the vector $\left[a_1, a_2, \ldots, a_{n-1}\right]^T$ and let $c = \frac{1}{1+a^Ta}$. Then it is easily verified that

$$\begin{bmatrix} 0 & 1 & 0 & \cdots & 0 \\ 0 & 0 & 1 & \ddots & \vdots \\ \vdots & & \ddots & \ddots & 0 \\ 0 & \cdots & \cdots & 0 & 1 \\ a_0 & a_1 & \cdots & & a_{n-1} \end{bmatrix}^+ = \begin{bmatrix} 0 & I \\ 0 & a^T \end{bmatrix}^+ = \begin{bmatrix} 0 & 0 \\ I - caa^T & ca \end{bmatrix}.$$

Note that $I - caa^T = \left(I + aa^T\right)^{-1}$, and hence the pseudoinverse of a singular companion matrix is not a companion matrix unless $a = 0$.

Companion matrices have many other interesting properties, among which, and perhaps surprisingly, is the fact that their singular values can be found in closed form; see [14].

Theorem 10.38. *Let $\sigma_1 \geq \sigma_2 \geq \cdots \geq \sigma_n$ be the singular values of the companion matrix A in (10.7). Let $\alpha = a_1^2 + a_2^2 + \cdots + a_{n-1}^2$ and $\gamma = 1 + a_0^2 + \alpha$. Then*

$$\begin{aligned} \sigma_1^2 &= \frac{1}{2}\left(\gamma + \sqrt{\gamma^2 - 4a_0^2}\right), \\ \sigma_i^2 &= 1 \quad \text{for } i = 2, 3, \ldots, n-1, \\ \sigma_n^2 &= \frac{1}{2}\left(\gamma - \sqrt{\gamma^2 - 4a_0^2}\right). \end{aligned}$$

If $a_0 \neq 0$, the largest and smallest singular values can also be written in the equivalent form

$$\sigma = \frac{2|a_0|}{\left|\sqrt{(1-a_0)^2 + \alpha} \mp \sqrt{(1+a_0)^2 + \alpha}\right|}.$$

Remark 10.39. Explicit formulas for all the associated right and left singular vectors can also be derived easily.

If $A \in \mathbb{R}^{n \times n}$ is **derogatory**, i.e., has more than one Jordan block associated with at least one eigenvalue, then it is not similar to a companion matrix of the form (10.7). However, it can be shown that a derogatory matrix is similar to a block diagonal matrix, each of whose diagonal blocks is a companion matrix. Such matrices are said to be in **rational canonical form** (or Frobenius canonical form). For details, see, for example, [12].

Companion matrices appear frequently in the control and signal processing literature but unfortunately they are often very difficult to work with numerically. Algorithms to reduce an arbitrary matrix to companion form are numerically unstable. Moreover, companion matrices are known to possess many undesirable numerical properties. For example, in general and especially as n increases, their eigenstructure is extremely ill conditioned, nonsingular ones are nearly singular, stable ones are nearly unstable, and so forth [14].

Companion matrices and rational canonical forms are generally to be avoided in floating-point computation.

Remark 10.40. Theorem 10.38 yields some understanding of why difficult numerical behavior might be expected for companion matrices. For example, when solving linear systems of equations of the form (6.2), one measure of numerical sensitivity is $\kappa_p(A) = \|A\|_p \|A^{-1}\|_p$, the so-called condition number of A with respect to inversion and with respect to the matrix p-norm. If this number is large, say $O(10^k)$, one may lose up to k digits of precision. In the 2-norm, this condition number is the ratio of largest to smallest singular values which, by the theorem, can be determined explicitly as

$$\frac{\sigma_1}{\sigma_n} = \frac{\gamma + \sqrt{\gamma^2 - 4a_0^2}}{2|a_0|}.$$

It is easy to show that $\frac{\gamma}{2|a_0|} \leq \kappa_2(A) \leq \frac{\gamma}{|a_0|}$, and when a_0 is small or γ is large (or both), then $\kappa_2(A) \approx \frac{\gamma}{|a_0|}$. It is not unusual for γ to be large for large n. Note that explicit formulas for $\kappa_1(A)$ and $\kappa_\infty(A)$ can also be determined easily by using (10.11).

EXERCISES

1. Show that if a triangular matrix is normal, then it must be diagonal.

2. Prove that if $A \in \mathbb{R}^{n \times n}$ is normal, then $\mathcal{N}(A) = \mathcal{N}(A^T)$.

3. Let $A \in \mathbb{C}^{n \times n}$ and define $\rho(A) = \max_{\lambda \in \Lambda(A)} |\lambda|$. Then $\rho(A)$ is called the **spectral radius** of A. Show that if A is normal, then $\rho(A) = \|A\|_2$. Show that the converse is true if $n = 2$.

4. Let $A \in \mathbb{C}^{n \times n}$ be normal with eigenvalues $\lambda_1, \ldots, \lambda_n$ and singular values $\sigma_1 \geq \sigma_2 \geq \cdots \geq \sigma_n \geq 0$. Show that $\sigma_i(A) = |\lambda_i(A)|$ for $i \in \underline{n}$.

5. Use the reverse-order identity matrix P introduced in (9.18) and the matrix U in Theorem 10.5 to find a unitary matrix Q that reduces $A \in \mathbb{C}^{n \times n}$ to lower triangular form.

6. Let $A = \begin{bmatrix} a & b \\ 0 & c \end{bmatrix} \in \mathbb{C}^{2 \times 2}$. Find a unitary matrix U such that

$$U^H A U = \begin{bmatrix} c & b \\ 0 & a \end{bmatrix}.$$

7. If $A \in \mathbb{R}^{n \times n}$ is positive definite, show that A^{-1} must also be positive definite.

8. Suppose $A \in \mathbb{R}^{n \times n}$ is positive definite. Is $\begin{bmatrix} A & I \\ I & A^{-1} \end{bmatrix} \geq 0$?

9. Let $R, S \in \mathbb{R}^{n \times n}$ be symmetric. Show that $\begin{bmatrix} R & I \\ I & S \end{bmatrix} > 0$ if and only if $S > 0$ and $R > S^{-1}$.

10. Find the inertia of the following matrices:

(a) $\begin{bmatrix} 0 & 1 \\ 1 & 0 \end{bmatrix}$, (b) $\begin{bmatrix} 1 & 1 \\ 1 & 1 \end{bmatrix}$, (c) $\begin{bmatrix} -2 & 1+j \\ 1-j & -2 \end{bmatrix}$,

(d) $\begin{bmatrix} -1 & 1+j \\ 1-j & -1 \end{bmatrix}$.

Chapter 11

Linear Differential and Difference Equations

11.1 Differential Equations

In this section we study solutions of the linear homogeneous system of differential equations

$$\dot{x}(t) = Ax(t); \quad x(t_0) = x_0 \in \mathbb{R}^n \tag{11.1}$$

for $t \geq t_0$. This is known as an **initial-value problem**. We restrict our attention in this chapter only to the so-called **time-invariant** case, where the matrix $A \in \mathbb{R}^{n \times n}$ is constant and does not depend on t. The solution of (11.1) is then known always to exist and be unique. It can be described conveniently in terms of the matrix exponential.

Definition 11.1. *For all $A \in \mathbb{R}^{n \times n}$, the* **matrix exponential** *$e^A \in \mathbb{R}^{n \times n}$ is defined by the power series*

$$e^A = \sum_{k=0}^{+\infty} \frac{1}{k!} A^k. \tag{11.2}$$

The series (11.2) can be shown to converge for all A (has radius of convergence equal to $+\infty$). The solution of (11.1) involves the matrix

$$e^{tA} = \sum_{k=0}^{+\infty} \frac{t^k}{k!} A^k, \tag{11.3}$$

which thus also converges for all A and uniformly in t.

11.1.1 Properties of the matrix exponential

1. $e^0 = I$.
   Proof: This follows immediately from Definition 11.1 by setting $A = 0$.

2. For all $A \in \mathbb{R}^{n \times n}$, $\left(e^A\right)^T = e^{A^T}$.
   Proof: This follows immediately from Definition 11.1 and linearity of the transpose.

3. For all $A \in \mathbb{R}^{n\times n}$ and for all $t, \tau \in \mathbb{R}$, $\quad e^{(t+\tau)A} = e^{tA}e^{\tau A} = e^{\tau A}e^{tA}$.
Proof: Note that

$$e^{(t+\tau)A} = I + (t+\tau)A + \frac{(t+\tau)^2}{2!}A^2 + \cdots$$

and

$$e^{tA}e^{\tau A} = \left(I + tA + \frac{t^2}{2!}A^2 + \cdots\right)\left(I + \tau A + \frac{\tau^2}{2!}A^2 + \cdots\right).$$

Compare like powers of A in the above two equations and use the binomial theorem on $(t+\tau)^k$.

4. For all $A, B \in \mathbb{R}^{n\times n}$ and for all $t \in \mathbb{R}$, $e^{t(A+B)} = e^{tA}e^{tB} = e^{tB}e^{tA}$ if and only if A and B **commute**, i.e., $AB = BA$.
Proof: Note that

$$e^{t(A+B)} = I + t(A+B) + \frac{t^2}{2!}(A+B)^2 + \cdots$$

and

$$e^{tA}e^{tB} = \left(I + tA + \frac{t^2}{2!}A^2 + \cdots\right)\left(I + tB + \frac{t^2}{2!}B^2 + \cdots\right)$$

while

$$e^{tB}e^{tA} = \left(I + tB + \frac{t^2}{2!}B^2 + \cdots\right)\left(I + tA + \frac{t^2}{2!}A^2 + \cdots\right).$$

Compare like powers of t in the first equation and the second or third and use the binomial theorem on $(A+B)^k$ and the commutativity of A and B.

5. For all $A \in \mathbb{R}^{n\times n}$ and for all $t \in \mathbb{R}$, $\quad (e^{tA})^{-1} = e^{-tA}$.
Proof: Simply take $\tau = -t$ in property 3.

6. Let $\mathcal{L}$ denote the Laplace transform and $\mathcal{L}^{-1}$ the inverse Laplace transform. Then for all $A \in \mathbb{R}^{n\times n}$ and for all $t \in \mathbb{R}$,

   (a) $\mathcal{L}\{e^{tA}\} = (sI - A)^{-1}$.

   (b) $\mathcal{L}^{-1}\{(sI - A)^{-1}\} = e^{tA}$.

   Proof: We prove only (a). Part (b) follows similarly.

$$\begin{aligned}
\mathcal{L}\{e^{tA}\} &= \int_0^{+\infty} e^{-st}e^{tA}\,dt \\
&= \int_0^{+\infty} e^{t(-sI)}e^{tA}\,dt \\
&= \int_0^{+\infty} e^{t(A-sI)}\,dt \quad \text{since } A \text{ and } (-sI) \text{ commute}
\end{aligned}$$

$$\begin{aligned}
&= \int_0^{+\infty} \sum_{i=1}^{n} e^{(\lambda_i - s)t} x_i y_i^H \, dt \quad \text{assuming } A \text{ is diagonalizable} \\
&= \sum_{i=1}^{n} \left[\int_0^{+\infty} e^{(\lambda_i - s)t} \, dt \right] x_i y_i^H \\
&= \sum_{i=1}^{n} \frac{1}{s - \lambda_i} x_i y_i^H \quad \text{assuming } \operatorname{Re} s > \operatorname{Re} \lambda_i \text{ for } i \in \underline{n} \\
&= (sI - A)^{-1}.
\end{aligned}$$

The matrix $(sI - A)^{-1}$ is called the **resolvent** of A and is defined for all s not in $\Lambda(A)$. Notice in the proof that we have assumed, for convenience, that A is diagonalizable. If this is not the case, the scalar dyadic decomposition can be replaced by

$$e^{t(A - sI)} = \sum_{i=1}^{m} X_i e^{t(J_i - sI)} Y_i^H$$

using the JCF. All succeeding steps in the proof then follow in a straightforward way.

7. For all $A \in \mathbb{R}^{n \times n}$ and for all $t \in \mathbb{R}$, $\frac{d}{dt}(e^{tA}) = Ae^{tA} = e^{tA}A$.
Proof: Since the series (11.3) is uniformly convergent, it can be differentiated term-by-term from which the result follows immediately. Alternatively, the formal definition

$$\frac{d}{dt}(e^{tA}) = \lim_{\Delta t \to 0} \frac{e^{(t+\Delta t)A} - e^{tA}}{\Delta t}$$

can be employed as follows. For any consistent matrix norm,

$$\begin{aligned}
\left\| \frac{e^{(t+\Delta t)A} - e^{tA}}{\Delta t} - Ae^{tA} \right\| &= \left\| \frac{1}{\Delta t}(e^{tA}e^{\Delta tA} - e^{tA}) - Ae^{tA} \right\| \\
&= \left\| \frac{1}{\Delta t}(e^{\Delta tA}e^{tA} - e^{tA}) - Ae^{tA} \right\| \\
&= \left\| \frac{1}{\Delta t}(e^{\Delta tA} - I)e^{tA} - Ae^{tA} \right\| \\
&= \left\| \frac{1}{\Delta t}\left(\Delta tA + \frac{(\Delta t)^2}{2!}A^2 + \cdots \right) e^{tA} - Ae^{tA} \right\| \\
&= \left\| \left(Ae^{tA} + \frac{\Delta t}{2!}A^2 e^{tA} + \cdots \right) - Ae^{tA} \right\| \\
&= \left\| \left(\frac{\Delta t}{2!}A^2 + \frac{(\Delta t)^2}{3!}A^3 + \cdots \right) e^{tA} \right\| \\
&\le \Delta t \|A^2\| \, \|e^{tA}\| \left(\frac{1}{2!} + \frac{\Delta t}{3!}\|A\| + \frac{(\Delta t)^2}{4!}\|A\|^2 + \cdots \right) \\
&< \Delta t \|A^2\| \, \|e^{tA}\| \left(1 + \Delta t\|A\| + \frac{(\Delta t)^2}{2!}\|A\|^2 + \cdots \right) \\
&= \Delta t \|A^2\| \, \|e^{tA}\| e^{\Delta t\|A\|}.
\end{aligned}$$

For fixed t, the right-hand side above clearly goes to 0 as Δt goes to 0. Thus, the limit exists and equals Ae^{tA}. A similar proof yields the limit $e^{tA}A$, or one can use the fact that A commutes with any polynomial of A of finite degree and hence with e^{tA}.

11.1.2 Homogeneous linear differential equations

Theorem 11.2. *Let $A \in \mathbb{R}^{n\times n}$. The solution of the linear homogeneous initial-value problem*

$$\dot{x}(t) = Ax(t); \quad x(t_0) = x_0 \in \mathbb{R}^n \tag{11.4}$$

for $t \geq t_0$ is given by

$$x(t) = e^{(t-t_0)A}x_0. \tag{11.5}$$

Proof: Differentiate (11.5) and use property 7 of the matrix exponential to get $\dot{x}(t) = Ae^{(t-t_0)A}x_0 = Ax(t)$. Also, $x(t_0) = e^{(t_0-t_0)A}x_0 = x_0$ so, by the fundamental existence and uniqueness theorem for ordinary differential equations, (11.5) is the solution of (11.4). □

11.1.3 Inhomogeneous linear differential equations

Theorem 11.3. *Let $A \in \mathbb{R}^{n\times n}$, $B \in \mathbb{R}^{n\times m}$ and let the vector-valued function u be given and, say, continuous. Then the solution of the linear inhomogeneous initial-value problem*

$$\dot{x}(t) = Ax(t) + Bu(t); \quad x(t_0) = x_0 \in \mathbb{R}^n \tag{11.6}$$

for $t \geq t_0$ is given by the **variation of parameters formula**

$$x(t) = e^{(t-t_0)A}x_0 + \int_{t_0}^{t} e^{(t-s)A}Bu(s)\,ds. \tag{11.7}$$

Proof: Differentiate (11.7) and again use property 7 of the matrix exponential. The general formula

$$\frac{d}{dt}\int_{p(t)}^{q(t)} f(x,t)\,dx = \int_{p(t)}^{q(t)} \frac{\partial f(x,t)}{\partial t}\,dx + f(q(t),t)\frac{dq(t)}{dt} - f(p(t),t)\frac{dp(t)}{dt}$$

is used to get $\dot{x}(t) = Ae^{(t-t_0)A}x_0 + \int_{t_0}^{t} Ae^{(t-s)A}Bu(s)\,ds + Bu(t) = Ax(t) + Bu(t)$. Also, $x(t_0) = e^{(t_0-t_0)A}x_0 + 0 = x_0$ so, by the fundamental existence and uniqueness theorem for ordinary differential equations, (11.7) is the solution of (11.6). □

Remark 11.4. The proof above simply verifies the variation of parameters formula by direct differentiation. The formula can be derived by means of an integrating factor "trick" as follows. Premultiply the equation $\dot{x} - Ax = Bu$ by e^{-tA} to get

$$e^{-tA}\dot{x} - e^{-tA}Ax = \frac{d}{dt}e^{-tA}x = e^{-tA}Bu. \tag{11.8}$$

Now integrate (11.8) over the interval $[t_0, t]$:

$$\int_{t_0}^{t} \frac{d}{ds} e^{-sA} x(s)\, ds = \int_{t_0}^{t} e^{-sA} Bu(s)\, ds.$$

Thus,

$$e^{-tA} x(t) - e^{-t_0 A} x(t_0) = \int_{t_0}^{t} e^{-sA} Bu(s)\, ds$$

and hence

$$x(t) = e^{(t-t_0)A} x_0 + \int_{t_0}^{t} e^{(t-s)A} Bu(s)\, ds.$$

11.1.4 Linear matrix differential equations

Matrix-valued initial-value problems also occur frequently. The first is an obvious generalization of Theorem 11.2, and the proof is essentially the same.

Theorem 11.5. *Let $A \in \mathbb{R}^{n \times n}$. The solution of the matrix linear homogeneous initial-value problem*

$$\dot{X}(t) = AX(t); \quad X(t_0) = C \in \mathbb{R}^{n \times n} \tag{11.9}$$

for $t \geq t_0$ is given by

$$X(t) = e^{(t-t_0)A} C. \tag{11.10}$$

In the matrix case, we can have coefficient matrices on both the right and left. For convenience, the following theorem is stated with initial time $t_0 = 0$.

Theorem 11.6. *Let $A \in \mathbb{R}^{n \times n}$, $B \in \mathbb{R}^{m \times m}$, and $C \in \mathbb{R}^{n \times m}$. Then the matrix initial-value problem*

$$\dot{X}(t) = AX(t) + X(t)B; \quad X(0) = C \tag{11.11}$$

has the solution $X(t) = e^{tA} C e^{tB}$.

Proof: Differentiate $e^{tA} C e^{tB}$ with respect to t and use property 7 of the matrix exponential. The fact that $X(t)$ satisfies the initial condition is trivial. □

Corollary 11.7. *Let $A, C \in \mathbb{R}^{n \times n}$. Then the matrix initial-value problem*

$$\dot{X}(t) = AX(t) + X(t)A^T; \quad X(0) = C \tag{11.12}$$

has the solution $X(t) = e^{tA} C e^{tA^T}$.

When C is symmetric in (11.12), $X(t)$ is symmetric and (11.12) is known as a **Lyapunov differential equation**. The initial-value problem (11.11) is known as a **Sylvester differential equation**.

11.1.5 Modal decompositions

Let $A \in \mathbb{R}^{n\times n}$ and suppose, for convenience, that it is diagonalizable (if A is not diagonalizable, the rest of this subsection is easily generalized by using the JCF and the decomposition $A = \sum X_i J_i Y_i^H$ as discussed in Chapter 9). Then the solution $x(t)$ of (11.4) can be written

$$\begin{aligned} x(t) &= e^{(t-t_0)A}x_0 \\ &= \left(\sum_{i=1}^{n} e^{\lambda_i(t-t_0)}x_i y_i^H\right)x_0 \\ &= \sum_{i=1}^{n}(y_i^H x_0 e^{\lambda_i(t-t_0)})x_i. \end{aligned}$$

The λ_is are called the **modal velocities** and the right eigenvectors x_i are called the **modal directions**. The decomposition above expresses the solution $x(t)$ as a weighted sum of its modal velocities and directions.

This modal decomposition can be expressed in a different looking but identical form if we write the initial condition x_0 as a weighted sum of the right eigenvectors $x_0 = \sum_{i=1}^{n}\alpha_i x_i$. Then

$$\begin{aligned} x(t) &= \left(\sum_{i=1}^{n} e^{\lambda_i(t-t_0)}x_i y_i^H\right)\left(\sum_{i=1}^{n}\alpha_i x_i\right) \\ &= \sum_{i=1}^{n}(\alpha_i e^{\lambda_i(t-t_0)})x_i. \end{aligned}$$

In the last equality we have used the fact that $y_i^H x_j = \delta_{ij}$.

Similarly, in the inhomogeneous case we can write

$$\int_{t_0}^{t} e^{(t-s)A}Bu(s)\,ds = \sum_{i=1}^{n}\left(\int_{t_0}^{t} e^{\lambda_i(t-s)}y_i^H Bu(s)\,ds\right)x_i.$$

11.1.6 Computation of the matrix exponential

JCF method

Let $A \in \mathbb{R}^{n\times n}$ and suppose $X \in \mathbb{R}_n^{n\times n}$ is such that $X^{-1}AX = J$, where J is a JCF for A. Then

$$\begin{aligned} e^{tA} &= e^{tXJX^{-1}} \\ &= Xe^{tJ}X^{-1} \\ &= \begin{cases} \sum_{i=1}^{n} e^{\lambda_i t}x_i y_i^H & \text{if } A \text{ is diagonalizable} \\ \sum_{i=1}^{m} X_i e^{tJ_i}Y_i^H & \text{in general.} \end{cases} \end{aligned}$$

If A is diagonalizable, it is then easy to compute e^{tA} via the formula $e^{tA} = Xe^{tJ}X^{-1}$ since e^{tJ} is simply a diagonal matrix.

In the more general case, the problem clearly reduces simply to the computation of the exponential of a Jordan block. To be specific, let $J_i \in \mathbb{C}^{k \times k}$ be a Jordan block of the form

$$J_i = \begin{bmatrix} \lambda & 1 & 0 & \cdots & 0 \\ 0 & \lambda & 1 & \ddots & \vdots \\ \vdots & \ddots & \lambda & \ddots & 0 \\ & & \ddots & \ddots & 1 \\ 0 & \cdots & & 0 & \lambda \end{bmatrix} = \lambda I + N.$$

Clearly λI and N commute. Thus, $e^{tJ_i} = e^{t\lambda I}e^{tN}$ by property 4 of the matrix exponential. The diagonal part is easy: $e^{t\lambda I} = \operatorname{diag}(e^{\lambda t}, \ldots, e^{\lambda t})$. But e^{tN} is almost as easy since N is nilpotent of degree k.

Definition 11.8. *A matrix $M \in \mathbb{R}^{n \times n}$ is* **nilpotent** *of degree (or index, or grade) p if $M^p = 0$, while $M^{p-1} \neq 0$.*

For the matrix N defined above, it is easy to check that while N has 1's along only its first superdiagonal (and 0's elsewhere), N^2 has 1's along only its second superdiagonal, and so forth. Finally, N^{k-1} has a 1 in its $(1, k)$ element and has 0's everywhere else, and $N^k = 0$. Thus, the series expansion of e^{tN} is finite, i.e.,

$$\begin{aligned} e^{tN} &= I + tN + \frac{t^2}{2!}N^2 + \cdots + \frac{t^{k-1}}{(k-1)!}N^{k-1} \\ &= \begin{bmatrix} 1 & t & \frac{t^2}{2!} & \cdots & \frac{t^{k-1}}{(k-1)!} \\ 0 & \ddots & \ddots & \ddots & \vdots \\ & \ddots & \ddots & \ddots & \frac{t^2}{2!} \\ \vdots & & \ddots & \ddots & t \\ 0 & \cdots & & 0 & 1 \end{bmatrix}. \end{aligned}$$

Thus,

$$e^{tJ_i} = \begin{bmatrix} e^{\lambda t} & te^{\lambda t} & \frac{t^2}{2!}e^{\lambda t} & \cdots & \frac{t^{k-1}}{(k-1)!}e^{\lambda t} \\ 0 & e^{\lambda t} & te^{\lambda t} & \ddots & \vdots \\ 0 & 0 & e^{\lambda t} & \ddots & \frac{t^2}{2!}e^{\lambda t} \\ \vdots & & \ddots & \ddots & te^{\lambda t} \\ 0 & \cdots & & 0 & e^{\lambda t} \end{bmatrix}.$$

In the case when λ is complex, a real version of the above can be worked out.

Example 11.9. Let $A = \left[\begin{smallmatrix} -4 & 4 \\ -1 & 0 \end{smallmatrix}\right]$. Then $\Lambda(A) = \{-2, -2\}$ and

$$\begin{aligned} e^{tA} &= Xe^{tJ}X^{-1} \\ &= \begin{bmatrix} 2 & 1 \\ 1 & 1 \end{bmatrix} \exp t \begin{bmatrix} -2 & 1 \\ 0 & -2 \end{bmatrix} \begin{bmatrix} 1 & -1 \\ -1 & 2 \end{bmatrix} \\ &= \begin{bmatrix} 2 & 1 \\ 1 & 1 \end{bmatrix} \begin{bmatrix} e^{-2t} & te^{-2t} \\ 0 & e^{-2t} \end{bmatrix} \begin{bmatrix} 1 & -1 \\ -1 & 2 \end{bmatrix} \\ &= \begin{bmatrix} e^{-2t} - 2te^{-2t} & 4te^{-2t} \\ -te^{-2t} & e^{-2t} + 2te^{-2t} \end{bmatrix}. \end{aligned}$$

Interpolation method

This method is numerically unstable in finite-precision arithmetic but is quite effective for hand calculation in small-order problems. The method is stated and illustrated for the exponential function but applies equally well to other functions.

Given $A \in \mathbb{R}^{n \times n}$ and $f(\lambda) = e^{t\lambda}$, compute $f(A) = e^{tA}$, where t is a fixed scalar. Suppose the characteristic polynomial of A can be written as $\pi(\lambda) = \prod_{i=1}^{m} (\lambda - \lambda_i)^{n_i}$, where the λ_is are distinct. Define

$$g(\lambda) = \alpha_0 + \alpha_1\lambda + \cdots + \alpha_{n-1}\lambda^{n-1},$$

where $\alpha_0, \ldots, \alpha_{n-1}$ are n constants that are to be determined. They are, in fact, the unique solution of the n equations:

$$g^{(k)}(\lambda_i) = f^{(k)}(\lambda_i); \quad k = 0, 1, \ldots, n_i - 1, \quad i \in \underline{m}.$$

Here, the superscript (k) denotes the kth derivative with respect to λ. With the α_is then known, the function g is known and $f(A) = g(A)$. The motivation for this method is the Cayley–Hamilton Theorem, Theorem 9.3, which says that all powers of A greater than $n - 1$ can be expressed as linear combinations of A^k for $k = 0, 1, \ldots, n - 1$. Thus, all the terms of order greater than $n - 1$ in the power series for e^{tA} can be written in terms of these lower-order powers as well. The polynomial g gives the appropriate linear combination.

Example 11.10. Let

$$A = \begin{bmatrix} -1 & 1 & 0 \\ 0 & -1 & 0 \\ 0 & 0 & -1 \end{bmatrix}$$

and $f(\lambda) = e^{t\lambda}$. Then $\pi(\lambda) = -(\lambda + 1)^3$, so $m = 1$ and $n_1 = 3$.

Let $g(\lambda) = \alpha_0 + \alpha_1\lambda + \alpha_2\lambda^2$. Then the three equations for the α_is are given by

$$\begin{aligned} g(-1) = f(-1) &\Longrightarrow \alpha_0 - \alpha_1 + \alpha_2 = e^{-t}, \\ g'(-1) = f'(-1) &\Longrightarrow \alpha_1 - 2\alpha_2 = te^{-t}, \\ g''(-1) = f''(-1) &\Longrightarrow 2\alpha_2 = t^2e^{-t}. \end{aligned}$$

Solving for the α_is, we find

$$\alpha_0 = e^{-t} + te^{-t} + \frac{t^2}{2}e^{-t},$$
$$\alpha_1 = te^{-t} + t^2e^{-t},$$
$$\alpha_2 = \frac{t^2}{2}e^{-t}.$$

Thus,

$$f(A) = e^{tA} = g(A) = \alpha_0 I + \alpha_1 A + \alpha_2 A^2 = \begin{bmatrix} e^{-t} & te^{-t} & 0 \\ 0 & e^{-t} & 0 \\ 0 & 0 & e^{-t} \end{bmatrix}.$$

Example 11.11. Let $A = \left[\begin{smallmatrix} -4 & 4 \\ -1 & 0 \end{smallmatrix}\right]$ and $f(\lambda) = e^{t\lambda}$. Then $\pi(\lambda) = (\lambda + 2)^2$ so $m = 1$ and $n_1 = 2$.

Let $g(\lambda) = \alpha_0 + \alpha_1\lambda$. Then the defining equations for the α_is are given by

$$g(-2) = f(-2) \Longrightarrow \alpha_0 - 2\alpha_1 = e^{-2t},$$
$$g'(-2) = f'(-2) \Longrightarrow \alpha_1 = te^{-2t}.$$

Solving for the α_is, we find

$$\alpha_0 = e^{-2t} + 2te^{-2t},$$
$$\alpha_1 = te^{-2t}.$$

Thus,

$$\begin{aligned} f(A) = e^{tA} = g(A) &= \alpha_0 I + \alpha_1 A \\ &= (e^{-2t} + 2te^{-2t})\begin{bmatrix} 1 & 0 \\ 0 & 1 \end{bmatrix} + te^{-2t}\begin{bmatrix} -4 & 4 \\ -1 & 0 \end{bmatrix} \\ &= \begin{bmatrix} e^{-2t} - 2te^{-2t} & 4te^{-2t} \\ -te^{-2t} & e^{-2t} + 2te^{-2t} \end{bmatrix}. \end{aligned}$$

Other methods

1. Use $e^{tA} = \mathcal{L}^{-1}\{(sI - A)^{-1}\}$ and techniques for inverse Laplace transforms. This is quite effective for small-order problems, but general nonsymbolic computational techniques are numerically unstable since the problem is theoretically equivalent to knowing precisely a JCF.

2. Use Padé approximation. There is an extensive literature on approximating certain nonlinear functions by rational functions. The matrix analogue yields $e^A \approx$

$D^{-1}(A)N(A)$, where $D(A) = \delta_0 I + \delta_1 A + \cdots + \delta_p A^p$ and $N(A) = \nu_0 I + \nu_1 A + \cdots + \nu_q A^q$. Explicit formulas are known for the coefficients of the numerator and denominator polynomials of various orders. Unfortunately, a Padé approximation for the exponential is accurate only in a neighborhood of the origin; in the matrix case this means when $\|A\|$ is sufficiently small. This can be arranged by scaling A, say, by multiplying it by $1/2^k$ for sufficiently large k and using the fact that $e^A = \left(e^{(1/2^k)A}\right)^{2^k}$. Numerical loss of accuracy can occur in this procedure from the successive squarings.

3. Reduce A to (real) Schur form S via the unitary similarity U and use $e^A = Ue^SU^H$ and successive recursions up the superdiagonals of the (quasi) upper triangular matrix e^S.

4. Many methods are outlined in, for example, [19]. Reliable and efficient computation of matrix functions such as e^A and $\log(A)$ remains a fertile area for research.

11.2 Difference Equations

In this section we outline solutions of discrete-time analogues of the linear differential equations of the previous section. Linear discrete-time systems, modeled by systems of difference equations, exhibit many parallels to the continuous-time differential equation case, and this observation is exploited frequently.

11.2.1 Homogeneous linear difference equations

Theorem 11.12. *Let $A \in \mathbb{R}^{n\times n}$. The solution of the linear homogeneous system of difference equations*

$$x_{k+1} = Ax_k; \quad x_0 \text{ given} \tag{11.13}$$

for $k \geq 0$ is given by

$$x_k = A^k x_0, \quad k \geq 0. \tag{11.14}$$

Proof: The proof is almost immediate upon substitution of (11.14) into (11.13). □

Remark 11.13. Again, we restrict our attention only to the so-called **time-invariant** case, where the matrix A in (11.13) is constant and does not depend on k. We could also consider an arbitrary "initial time" k_0, but since the system is time-invariant, and since we want to keep the formulas "clean" (i.e., no double subscripts), we have chosen $k_0 = 0$ for convenience.

11.2.2 Inhomogeneous linear difference equations

Theorem 11.14. *Let $A \in \mathbb{R}^{n\times n}$, $B \in \mathbb{R}^{n\times m}$ and suppose $\{u_k\}_{k=0}^{+\infty}$ is a given sequence of m-vectors. Then the solution of the inhomogeneous initial-value problem*

$$x_{k+1} = Ax_k + Bu_k; \quad x_0 \text{ given} \tag{11.15}$$

is given by

$$x_k = A^k x_0 + \sum_{j=0}^{k-1} A^{k-j-1} B u_j, \quad k \geq 0. \tag{11.16}$$

Proof: The proof is again almost immediate upon substitution of (11.16) into (11.15). □

11.2.3 Computation of matrix powers

It is clear that solution of linear systems of difference equations involves computation of A^k. One solution method, which is numerically unstable but sometimes useful for hand calculation, is to use z-transforms, by analogy with the use of Laplace transforms to compute a matrix exponential. One definition of the z-transform of a sequence $\{g_k\}$ is

$$\mathcal{Z}(\{g_k\}_{k=0}^{+\infty}) = \sum_{k=0}^{+\infty} g_k z^{-k}.$$

Assuming $|z| > \max_{\lambda \in \Lambda(A)} |\lambda|$, the z-transform of the sequence $\{A^k\}$ is then given by

$$\begin{aligned} \mathcal{Z}(\{A^k\}) = \sum_{k=0}^{+\infty} z^{-k} A^k &= I + \frac{1}{z}A + \frac{1}{z^2}A^2 + \cdots \\ &= (I - z^{-1}A)^{-1} \\ &= z(zI - A)^{-1}. \end{aligned}$$

Methods based on the JCF are sometimes useful, again mostly for small-order problems. Assume that $A \in \mathbb{R}^{n \times n}$ and let $X \in \mathbb{R}_n^{n \times n}$ be such that $X^{-1}AX = J$, where J is a JCF for A. Then

$$\begin{aligned} A^k &= (XJX^{-1})^k \\ &= XJ^kX^{-1} \\ &= \begin{cases} \sum_{i=1}^{n} \lambda_i^k x_i y_i^H & \text{if } A \text{ is diagonalizable,} \\ \sum_{i=1}^{m} X_i J_i^k Y_i^H & \text{in general.} \end{cases} \end{aligned}$$

If A is diagonalizable, it is then easy to compute A^k via the formula $A^k = XJ^kX^{-1}$ since J^k is simply a diagonal matrix.

In the general case, the problem again reduces to the computation of the power of a Jordan block. To be specific, let $J_i \in \mathbb{C}^{p\times p}$ be a Jordan block of the form

$$J_i = \begin{bmatrix} \lambda & 1 & 0 & \cdots & 0 \\ 0 & \lambda & 1 & \ddots & \vdots \\ \vdots & \ddots & \lambda & \ddots & 0 \\ & & \ddots & \ddots & 1 \\ 0 & \cdots & & 0 & \lambda \end{bmatrix}.$$

Writing $J_i = \lambda I + N$ and noting that λI and the nilpotent matrix N commute, it is then straightforward to apply the binomial theorem to $(\lambda I + N)^k$ and verify that

$$J_i^k = \begin{bmatrix} \lambda^k & k\lambda^{k-1} & \binom{k}{2}\lambda^{k-2} & \cdots & \binom{k}{p-1}\lambda^{k-p+1} \\ 0 & \lambda^k & k\lambda^{k-1} & \ddots & \vdots \\ 0 & 0 & \lambda^k & \ddots & \binom{k}{2}\lambda^{k-2} \\ \vdots & & \ddots & \ddots & k\lambda^{k-1} \\ 0 & \cdots & & 0 & \lambda^k \end{bmatrix}.$$

The symbol $\binom{k}{q}$ has the usual definition of $\frac{k!}{q!(k-q)!}$ and is to be interpreted as 0 if $k < q$.

In the case when λ is complex, a real version of the above can be worked out.

Example 11.15. Let $A = \left[\begin{smallmatrix} -4 & 4 \\ -1 & 0 \end{smallmatrix}\right]$. Then

$$\begin{aligned} A^k = XJ^kX^{-1} &= \begin{bmatrix} 2 & 1 \\ 1 & 1 \end{bmatrix}\begin{bmatrix} (-2)^k & k(-2)^{k-1} \\ 0 & (-2)^k \end{bmatrix}\begin{bmatrix} 1 & -1 \\ -1 & 2 \end{bmatrix} \\ &= \begin{bmatrix} (-2)^{k-1}(-2-2k) & k(-2)^{k+1} \\ -k(-2)^{k-1} & (-2)^{k-1}(2k-2) \end{bmatrix}. \end{aligned}$$

Basic analogues of other methods such as those mentioned in Section 11.1.6 can also be derived for the computation of matrix powers, but again no universally "best" method exists. For an erudite discussion of the state of the art, see [11, Ch. 18].

11.3 Higher-Order Equations

It is well known that a higher-order (scalar) linear differential equation can be converted to a first-order linear system. Consider, for example, the initial-value problem

$$y^{(n)}(t) + a_{n-1}y^{(n-1)}(t) + \cdots + a_1\dot{y}(t) + a_0y(t) = \phi(t) \tag{11.17}$$

with $\phi(t)$ a given function and n initial conditions

$$y(0) = c_0, \;\; \dot{y}(0) = c_1, \;\; \ldots, \;\; y^{(n-1)}(0) = c_{n-1}. \tag{11.18}$$

Here, $y^{(m)}$ denotes the mth derivative of y with respect to t. Define a vector $x(t) \in \mathbb{R}^n$ with components $x_1(t) = y(t),\ x_2(t) = \dot{y}(t),\ \ldots,\ x_n(t) = y^{(n-1)}(t)$. Then

$$\begin{aligned}
\dot{x}_1(t) &= x_2(t) = \dot{y}(t), \\
\dot{x}_2(t) &= x_3(t) = \ddot{y}(t), \\
&\ \ \vdots \qquad \vdots \\
\dot{x}_{n-1}(t) &= x_n(t) = y^{(n-1)}(t), \\
\dot{x}_n(t) &= y^{(n)}(t) = -a_0 y(t) - a_1 \dot{y}(t) - \cdots - a_{n-1} y^{(n-1)}(t) + \phi(t) \\
&= -a_0 x_1(t) - a_1 x_2(t) - \cdots - a_{n-1} x_n(t) + \phi(t).
\end{aligned}$$

These equations can then be rewritten as the first-order linear system

$$\dot{x}(t) = \begin{bmatrix} 0 & 1 & 0 & \cdots & 0 \\ 0 & 0 & 1 & \ddots & \vdots \\ \vdots & & \ddots & \ddots & 0 \\ 0 & \cdots & \cdots & 0 & 1 \\ -a_0 & -a_1 & \cdots & & -a_{n-1} \end{bmatrix} x(t) + \begin{bmatrix} 0 \\ \vdots \\ 0 \\ 1 \end{bmatrix} \phi(t). \tag{11.19}$$

The initial conditions take the form $x(0) = c = \left[c_0, c_1, \ldots, c_{n-1}\right]^T$.

Note that $\det(\lambda I - A) = \lambda^n + a_{n-1}\lambda^{n-1} + \cdots + a_1\lambda + a_0$. However, the companion matrix A in (11.19) possesses many nasty numerical properties for even moderately sized n and, as mentioned before, is often well worth avoiding, at least for computational purposes.

A similar procedure holds for the conversion of a higher-order difference equation

$$y_{k+n} + a_{n-1} y_{k+n-1} + \cdots + a_1 y_{k+1} + a_0 y_k = \phi_k,$$

with n initial conditions, into a linear first-order difference equation with (vector) initial condition.

EXERCISES

1. Let $P \in \mathbb{R}^{n\times n}$ be a projection. Show that $e^P \approx I + 1.718P$.

2. Suppose $x, y \in \mathbb{R}^n$ and let $A = xy^T$. Further, let $\alpha = x^T y$. Show that $e^{tA} = I + g(t, \alpha) xy^T$, where

$$g(t, \alpha) = \begin{cases} \frac{1}{\alpha}(e^{\alpha t} - 1) & \text{if } \alpha \neq 0, \\ t & \text{if } \alpha = 0. \end{cases}$$

3. Let

$$A = \begin{bmatrix} I & X \\ 0 & -I \end{bmatrix},$$

where $X \in \mathbb{R}^{m \times n}$ is arbitrary. Show that

$$e^A = \begin{bmatrix} eI & \sinh 1\, X \\ 0 & \frac{1}{e} I \end{bmatrix}.$$

4. Let K denote the skew-symmetric matrix

$$\begin{bmatrix} 0 & I_n \\ -I_n & 0 \end{bmatrix},$$

where I_n denotes the $n \times n$ identity matrix. A matrix $A \in \mathbb{R}^{2n \times 2n}$ is said to be **Hamiltonian** if $K^{-1} A^T K = -A$ and to be **symplectic** if $K^{-1} A^T K = A^{-1}$.

(a) Suppose H is Hamiltonian and let λ be an eigenvalue of H. Show that $-\lambda$ must also be an eigenvalue of H.

(b) Suppose S is symplectic and let λ be an eigenvalue of S. Show that $1/\lambda$ must also be an eigenvalue of S.

(c) Suppose that H is Hamiltonian and S is symplectic. Show that $S^{-1} H S$ must be Hamiltonian.

(d) Suppose H is Hamiltonian. Show that e^H must be symplectic.

5. Let $\alpha, \beta \in \mathbb{R}$ and

$$A = \begin{bmatrix} \alpha & \beta \\ -\beta & \alpha \end{bmatrix}.$$

Then show that

$$e^{tA} = \begin{bmatrix} e^{\alpha t} \cos \beta t & e^{\alpha t} \sin \beta t \\ -e^{\alpha t} \sin \beta t & e^{\alpha t} \cos \beta t \end{bmatrix}.$$

6. Find a general expression for

$$\begin{bmatrix} \alpha & \beta \\ -\beta & \alpha \end{bmatrix}^k.$$

7. Find e^{tA} when $A =$

$$\text{(a)} \begin{bmatrix} 2 & 1 \\ 1 & 2 \end{bmatrix}, \quad \text{(b)} \begin{bmatrix} 1 & 1 \\ -1 & 3 \end{bmatrix}, \quad \text{(c)} \begin{bmatrix} -2 & 1 \\ -1 & -2 \end{bmatrix}.$$

8. Let

$$A = \begin{bmatrix} -1 & 1 \\ 0 & -1 \end{bmatrix}, \quad b = \begin{bmatrix} 2 \\ 3 \end{bmatrix}.$$

(a) Solve the differential equation

$$\dot{x} = Ax \; ; \quad x(0) = \begin{bmatrix} 1 \\ 2 \end{bmatrix}.$$

(b) Solve the differential equation

$$\dot{x} = Ax + b\,; \qquad x(0) = \begin{bmatrix} 1 \\ 2 \end{bmatrix}.$$

9. Consider the initial-value problem

$$\dot{x}(t) = Ax(t); \qquad x(0) = x_0$$

for $t \geq 0$. Suppose that $A \in \mathbb{R}^{n \times n}$ is skew-symmetric and let $\alpha = \|x_0\|_2$. Show that $\|x(t)\|_2 = \alpha$ for all $t > 0$.

10. Consider the $n \times n$ matrix initial-value problem

$$\dot{X}(t) = AX(t) - X(t)A; \qquad X(0) = C.$$

Show that the eigenvalues of the solution $X(t)$ of this problem are the same as those of C for all t.

11. The year is 2004 and there are three large "free trade zones" in the world: Asia (A), Europe (E), and the Americas (R). Suppose certain multinational companies have total assets of \$40 trillion of which \$20 trillion is in E and \$20 trillion is in R. Each year half of the Americas' money stays home, a quarter goes to Europe, and a quarter goes to Asia. For Europe and Asia, half stays home and half goes to the Americas.

(a) Find the matrix M that gives

$$\begin{bmatrix} \text{A} \\ \text{E} \\ \text{R} \end{bmatrix}_{\text{year } k+1} = M \begin{bmatrix} \text{A} \\ \text{E} \\ \text{R} \end{bmatrix}_{\text{year } k}.$$

(b) Find the eigenvalues and right eigenvectors of M.

(c) Find the distribution of the companies' assets at year k.

(d) Find the limiting distribution of the \$40 trillion as the universe ends, i.e., as $k \to +\infty$ (i.e., around the time the Cubs win a World Series).

(Exercise adapted from Problem 5.3.11 in [24].)

12. (a) Find the solution of the initial-value problem

$$\ddot{y}(t) + 2\dot{y}(t) + y(t) = 0; \qquad y(0) = 1, \; \dot{y}(0) = 0.$$

(b) Consider the difference equation

$$z_{k+2} + 2z_{k+1} + z_k = 0.$$

If $z_0 = 1$ and $z_1 = 2$, what is the value of z_{1000}? What is the value of z_k in general?

Chapter 12

Generalized Eigenvalue Problems

12.1 The Generalized Eigenvalue/Eigenvector Problem

In this chapter we consider the **generalized eigenvalue problem**

$$Ax = \lambda Bx,$$

where $A, B \in \mathbb{C}^{n \times n}$. The standard eigenvalue problem considered in Chapter 9 obviously corresponds to the special case that $B = I$.

Definition 12.1. *A nonzero vector $x \in \mathbb{C}^n$ is a* **right generalized eigenvector** *of the pair (A, B) with $A, B \in \mathbb{C}^{n \times n}$ if there exists a scalar $\lambda \in \mathbb{C}$, called a* **generalized eigenvalue**, *such that*

$$Ax = \lambda Bx. \tag{12.1}$$

Similarly, a nonzero vector $y \in \mathbb{C}^n$ is a **left generalized eigenvector** *corresponding to an eigenvalue λ if*

$$y^H A = \lambda y^H B. \tag{12.2}$$

When the context is such that no confusion can arise, the adjective "generalized" is usually dropped. As with the standard eigenvalue problem, if x [y] is a right [left] eigenvector, then so is αx [αy] for any nonzero scalar $\alpha \in \mathbb{C}$.

Definition 12.2. *The matrix $A - \lambda B$ is called a* **matrix pencil** *(or pencil of the matrices A and B).*

As with the standard eigenvalue problem, eigenvalues for the generalized eigenvalue problem occur where the matrix pencil $A - \lambda B$ is singular.

Definition 12.3. *The polynomial $\pi(\lambda) = \det(A - \lambda B)$ is called the* **characteristic polynomial** *of the matrix pair (A, B). The roots of $\pi(\lambda)$ are the eigenvalues of the associated generalized eigenvalue problem.*

Remark 12.4. When $A, B \in \mathbb{R}^{n \times n}$, the characteristic polynomial is obviously real, and hence nonreal eigenvalues must occur in complex conjugate pairs.

Remark 12.5. If $B = I$ (or in general when B is nonsingular), then $\pi(\lambda)$ is a polynomial of degree n, and hence there are n eigenvalues associated with the pencil $A - \lambda B$. However, when $B \neq I$, in particular, when B is singular, there may be 0, $k \in \underline{n}$, or infinitely many eigenvalues associated with the pencil $A - \lambda B$. For example, suppose

$$A = \begin{bmatrix} 1 & 0 \\ 0 & \alpha \end{bmatrix}, \quad B = \begin{bmatrix} 1 & 0 \\ 0 & \beta \end{bmatrix}, \tag{12.3}$$

where α and β are scalars. Then the characteristic polynomial is

$$\det(A - \lambda B) = (1 - \lambda)(\alpha - \beta\lambda)$$

and there are several cases to consider.

Case 1: $\alpha \neq 0$, $\beta \neq 0$. There are two eigenvalues, 1 and $\frac{\alpha}{\beta}$.

Case 2: $\alpha = 0$, $\beta \neq 0$. There are two eigenvalues, 1 and 0.

Case 3: $\alpha \neq 0$, $\beta = 0$. There is only one eigenvalue, 1 (of multiplicity 1).

Case 4: $\alpha = 0$, $\beta = 0$. All $\lambda \in \mathbb{C}$ are eigenvalues since $\det(A - \lambda B) \equiv 0$.

Definition 12.6. *If* $\det(A - \lambda B)$ *is not identically zero, the pencil* $A - \lambda B$ *is said to be* **regular**; *otherwise, it is said to be* **singular**.

Note that if $\mathcal{N}(A) \cap \mathcal{N}(B) \neq 0$, the associated matrix pencil is singular (as in Case 4 above).

Associated with any matrix pencil $A - \lambda B$ is a **reciprocal** pencil $B - \mu A$ and corresponding generalized eigenvalue problem. Clearly the reciprocal pencil has eigenvalues $\mu = \frac{1}{\lambda}$. It is instructive to consider the reciprocal pencil associated with the example in Remark 12.5. With A and B as in (12.3), the characteristic polynomial is

$$\det(B - \mu A) = (1 - \mu)(\beta - \alpha\mu)$$

and there are again four cases to consider.

Case 1: $\alpha \neq 0$, $\beta \neq 0$. There are two eigenvalues, 1 and $\frac{\beta}{\alpha}$.

Case 2: $\alpha = 0$, $\beta \neq 0$. There is only one eigenvalue, 1 (of multiplicity 1).

Case 3: $\alpha \neq 0$, $\beta = 0$. There are two eigenvalues, 1 and 0.

Case 4: $\alpha = 0$, $\beta = 0$. All $\lambda \in \mathbb{C}$ are eigenvalues since $\det(B - \mu A) \equiv 0$.

At least for the case of regular pencils, it is apparent where the "missing" eigenvalues have gone in Cases 2 and 3. That is to say, there is a second eigenvalue "at infinity" for Case 3 of $A - \lambda B$, with its reciprocal eigenvalue being 0 in Case 3 of the reciprocal pencil $B - \mu A$. A similar reciprocal symmetry holds for Case 2.

While there are applications in system theory and control where singular pencils appear, only the case of regular pencils is considered in the remainder of this chapter. Note that A and/or B may still be singular. If B is singular, the pencil $A - \lambda B$ always has fewer than n eigenvalues. If B is nonsingular, the pencil $A - \lambda B$ always has precisely n eigenvalues, since the generalized eigenvalue problem is then easily seen to be equivalent to the standard eigenvalue problem $B^{-1}Ax = \lambda x$ (or $AB^{-1}w = \lambda w$). However, this turns out to be a very poor numerical procedure for handling the generalized eigenvalue problem if B is even moderately ill conditioned with respect to inversion. Numerical methods that work directly on A and B are discussed in standard textbooks on numerical linear algebra; see, for example, [7, Sec. 7.7] or [25, Sec. 6.7].

12.2 Canonical Forms

Just as for the standard eigenvalue problem, canonical forms are available for the generalized eigenvalue problem. Since the latter involves a pair of matrices, we now deal with equivalencies rather than similarities, and the first theorem deals with what happens to eigenvalues and eigenvectors under equivalence.

Theorem 12.7. *Let $A, B, Q, Z \in \mathbb{C}^{n\times n}$ with Q and Z nonsingular. Then*

1. *the eigenvalues of the problems $A - \lambda B$ and $QAZ - \lambda QBZ$ are the same (the two problems are said to be* **equivalent**).
2. *if x is a right eigenvector of $A - \lambda B$, then $Z^{-1}x$ is a right eigenvector of $QAZ - \lambda QBZ$.*
3. *if y is a left eigenvector of $A - \lambda B$, then $Q^{-H}y$ is a left eigenvector of $QAZ - \lambda QBZ$.*

Proof:

1. $\det(QAZ - \lambda QBZ) = \det[Q(A - \lambda B)Z] = \det Q \det Z \det(A - \lambda B)$. Since $\det Q$ and $\det Z$ are nonzero, the result follows.
2. The result follows by noting that $(A - \lambda B)x = 0$ if and only if $Q(A - \lambda B)Z(Z^{-1}x) = 0$.
3. Again, the result follows easily by noting that $y^H(A - \lambda B) = 0$ if and only if $(Q^{-H}y)^H Q(A - \lambda B)Z = 0$. □

The first canonical form is an analogue of Schur's Theorem and forms, in fact, the theoretical foundation for the QZ algorithm, which is the generally preferred method for solving the generalized eigenvalue problem; see, for example, [7, Sec. 7.7] or [25, Sec. 6.7].

Theorem 12.8. *Let $A, B \in \mathbb{C}^{n\times n}$. Then there exist unitary matrices $Q, Z \in \mathbb{C}^{n\times n}$ such that*

$$QAZ = T_\alpha, \quad QBZ = T_\beta,$$

where T_α and T_β are upper triangular.

By Theorem 12.7, the eigenvalues of the pencil $A - \lambda B$ are then the ratios of the diagonal elements of T_α to the corresponding diagonal elements of T_β, with the understanding that a zero diagonal element of T_β corresponds to an infinite generalized eigenvalue.

There is also an analogue of the Murnaghan–Wintner Theorem for real matrices.

Theorem 12.9. *Let $A, B \in \mathbb{R}^{n \times n}$. Then there exist orthogonal matrices $Q, Z \in \mathbb{R}^{n \times n}$ such that*

$$QAZ = S, \quad QBZ = T,$$

where T is upper triangular and S is quasi-upper-triangular.

When S has a 2×2 diagonal block, the 2×2 subpencil formed with the corresponding 2×2 diagonal subblock of T has a pair of complex conjugate eigenvalues. Otherwise, real eigenvalues are given as above by the ratios of diagonal elements of S to corresponding elements of T.

There is also an analogue of the Jordan canonical form called the **Kronecker canonical form** (KCF). A full description of the KCF, including analogues of principal vectors and so forth, is beyond the scope of this book. In this chapter, we present only statements of the basic theorems and some examples. The first theorem pertains only to "square" regular pencils, while the full KCF in all its generality applies also to "rectangular" and singular pencils.

Theorem 12.10. *Let $A, B \in \mathbb{C}^{n \times n}$ and suppose the pencil $A - \lambda B$ is regular. Then there exist nonsingular matrices $P, Q \in \mathbb{C}^{n \times n}$ such that*

$$P(A - \lambda B)Q = \begin{bmatrix} J & 0 \\ 0 & I \end{bmatrix} - \lambda \begin{bmatrix} I & 0 \\ 0 & N \end{bmatrix},$$

where J is a Jordan canonical form corresponding to the finite eigenvalues of $A - \lambda B$ and N is a nilpotent matrix of Jordan blocks associated with 0 and corresponding to the infinite eigenvalues of $A - \lambda B$.

Example 12.11. The matrix pencil

$$\begin{bmatrix} 2 & 1 & 0 & 0 & 0 \\ 0 & 2 & 0 & 0 & 0 \\ 0 & 0 & 1 & 0 & 0 \\ 0 & 0 & 0 & 1 & 0 \\ 0 & 0 & 0 & 0 & 1 \end{bmatrix} - \lambda \begin{bmatrix} 1 & 0 & 0 & 0 & 0 \\ 0 & 1 & 0 & 0 & 0 \\ 0 & 0 & 0 & 1 & 0 \\ 0 & 0 & 0 & 0 & 0 \\ 0 & 0 & 0 & 0 & 0 \end{bmatrix}$$

with characteristic polynomial $(\lambda - 2)^2$ has a finite eigenvalue 2 of multiplicty 2 and three infinite eigenvalues.

Theorem 12.12 (Kronecker Canonical Form). *Let $A, B \in \mathbb{C}^{m \times n}$. Then there exist nonsingular matrices $P \in \mathbb{C}^{m \times m}$ and $Q \in \mathbb{C}^{n \times n}$ such that*

$$P(A - \lambda B)Q = \operatorname{diag}(L_{l_1}, \ldots, L_{l_s}, L_{r_1}^T, \ldots, L_{r_t}^T, J - \lambda I, I - \lambda N),$$

where N is nilpotent, both N and J are in Jordan canonical form, and L_k is the $(k+1) \times k$ bidiagonal pencil

$$L_k = \begin{bmatrix} -\lambda & 0 & \cdots & 0 \\ 1 & -\lambda & \ddots & \vdots \\ 0 & \ddots & \ddots & 0 \\ \vdots & \ddots & \ddots & -\lambda \\ 0 & \cdots & 0 & 1 \end{bmatrix}.$$

The l_i are called the left minimal indices while the r_i are called the right minimal indices. Left or right minimal indices can take the value 0.

Example 12.13. Consider a 13×12 block diagonal matrix whose diagonal blocks are

$$\begin{bmatrix} 0 & 0 \\ 0 & 0 \\ 0 & 0 \end{bmatrix}, \begin{bmatrix} -\lambda \\ 1 \end{bmatrix}, \begin{bmatrix} -\lambda & 1 & 0 \\ 0 & -\lambda & 1 \end{bmatrix}, \begin{bmatrix} 2-\lambda & 1 \\ 0 & 2-\lambda \end{bmatrix}, [3-\lambda], \begin{bmatrix} 1 & -\lambda & 0 \\ 0 & 1 & -\lambda \\ 0 & 0 & 1 \end{bmatrix}.$$

Such a matrix is in KCF. The first block of zeros actually corresponds to L_0, L_0, L_0, L_0^T, L_0^T, where each L_0 has "zero columns" and one row, while each L_0^T has "zero rows" and one column. The second block is L_1 while the third block is L_2^T. The next two blocks correspond to

$$J = \begin{bmatrix} 2 & 1 & 0 \\ 0 & 2 & 0 \\ 0 & 0 & 3 \end{bmatrix}$$

while the nilpotent matrix N in this example is

$$\begin{bmatrix} 0 & 1 & 0 \\ 0 & 0 & 1 \\ 0 & 0 & 0 \end{bmatrix}.$$

Just as sets of eigenvectors span A-invariant subspaces in the case of the standard eigenproblem (recall Definition 9.35), there is an analogous geometric concept for the generalized eigenproblem.

Definition 12.14. *Let $A, B \in \mathbb{R}^{n \times n}$ and suppose the pencil $A - \lambda B$ is regular. Then $\mathcal{V}$ is a* **deflating subspace** *if*

$$\dim(A\mathcal{V} + B\mathcal{V}) = \dim \mathcal{V}. \tag{12.4}$$

Just as in the standard eigenvalue case, there is a matrix characterization of deflating subspace. Specifically, suppose $S \in \mathbb{R}^{n \times k}$ is a matrix whose columns span a k-dimensional subspace $\mathcal{S}$ of $\mathbb{R}^n$, i.e., $\mathcal{R}(S) = \mathcal{S}$. Then $\mathcal{S}$ is a deflating subspace for the pencil $A - \lambda B$ if and only if there exists $M \in \mathbb{R}^{k \times k}$ such that

$$AS = BSM. \tag{12.5}$$

If $B = I$, then (12.4) becomes $\dim(A\mathcal{V} + \mathcal{V}) = \dim\mathcal{V}$, which is clearly equivalent to $A\mathcal{V} \subseteq \mathcal{V}$. Similarly, (12.5) becomes $AS = SM$ as before. If the pencil is not regular, there is a concept analogous to deflating subspace called a **reducing subspace**.

12.3 Application to the Computation of System Zeros

Consider the linear system

$$\dot{x} = Ax + Bu,$$
$$y = Cx + Du$$

with $A \in \mathbb{R}^{n\times n}$, $B \in \mathbb{R}^{n\times m}$, $C \in \mathbb{R}^{p\times n}$, and $D \in \mathbb{R}^{p\times m}$. This linear time-invariant state-space model is often used in multivariable control theory, where $x(= x(t))$ is called the state vector, u is the vector of inputs or controls, and y is the vector of outputs or observables. For details, see, for example, [26].

In general, the (finite) zeros of this system are given by the (finite) complex numbers z, where the "system pencil"

$$\begin{bmatrix} A & B \\ C & D \end{bmatrix} - \lambda \begin{bmatrix} I & 0 \\ 0 & 0 \end{bmatrix} \tag{12.6}$$

drops rank. In the special case $p = m$, these values are the generalized eigenvalues of the $(n + m) \times (n + m)$ pencil.

Example 12.15. Let

$$A = \begin{bmatrix} -4 & -3 \\ 2 & 1 \end{bmatrix}, \quad B = \begin{bmatrix} 3 \\ 1 \end{bmatrix}, \quad C = [1 \;\; 2], \quad D = 0.$$

Then the transfer matrix (see [26]) of this system is

$$g(s) = C(sI - A)^{-1}B + D = \frac{5s + 14}{s^2 + 3s + 2},$$

which clearly has a zero at -2.8. Checking the finite eigenvalues of the pencil (12.6), we find the characteristic polynomial to be

$$\det \begin{bmatrix} A - \lambda I & B \\ C & D \end{bmatrix} = 5\lambda + 14,$$

which has a root at -2.8.

The method of finding system zeros via a generalized eigenvalue problem also works well for general multi-input, multi-output systems. Numerically, however, one must be careful first to "deflate out" the infinite zeros (infinite eigenvalues of (12.6)). This is accomplished by computing a certain unitary equivalence on the system pencil that then yields a smaller generalized eigenvalue problem with only finite generalized eigenvalues (the finite zeros).

The connection between system zeros and the corresponding system pencil is nontrivial. However, we offer some insight below into the special case of a single-input, single-output system. Specifically, let $B = b \in \mathbb{R}^n$, $C = c^T \in \mathbb{R}^{1\times n}$, and $D = d \in \mathbb{R}$. Furthermore, let $g(s) = c^T(sI - A)^{-1}b + d$ denote the system transfer function (matrix), and assume that $g(s)$ can be written in the form

$$g(s) = \frac{\nu(s)}{\pi(s)},$$

where $\pi(s)$ is the characteristic polynomial of A, and $\nu(s)$ and $\pi(s)$ are relatively prime (i.e., there are no "pole/zero cancellations").

Suppose $z \in \mathbb{C}$ is such that

$$\begin{bmatrix} A - zI & b \\ c^T & d \end{bmatrix}$$

is singular. Then there exists a nonzero solution to

$$\begin{bmatrix} A - zI & b \\ c^T & d \end{bmatrix} \begin{bmatrix} x \\ y \end{bmatrix} = 0$$

or

$$(A - zI)x + by = 0, \tag{12.7}$$

$$c^T x + dy = 0. \tag{12.8}$$

Assuming z is not an eigenvalue of A (i.e., no pole/zero cancellations), then from (12.7) we get

$$x = -(A - zI)^{-1}by. \tag{12.9}$$

Substituting this in (12.8), we have

$$-c^T(A - zI)^{-1}by + dy = 0,$$

or $g(z)y = 0$ by the definition of g. Now $y \neq 0$ (else $x = 0$ from (12.9)). Hence $g(z) = 0$, i.e., z is a zero of g.

12.4 Symmetric Generalized Eigenvalue Problems

A very important special case of the generalized eigenvalue problem

$$Ax = \lambda Bx \tag{12.10}$$

for $A, B \in \mathbb{R}^{n\times n}$ arises when $A = A^T$ and $B = B^T > 0$. For example, the second-order system of differential equations

$$M\ddot{x} + Kx = 0,$$

where M is a symmetric positive definite "mass matrix" and K is a symmetric "stiffness matrix," is a frequently employed model of structures or vibrating systems and yields a generalized eigenvalue problem of the form (12.10).

Since B is positive definite it is nonsingular. Thus, the problem (12.10) is equivalent to the standard eigenvalue problem $B^{-1}Ax = \lambda x$. However, $B^{-1}A$ is not necessarily symmetric.

Example 12.16. Let $A = \begin{bmatrix} 1 & 3 \\ 3 & 2 \end{bmatrix}$, $B = \begin{bmatrix} 2 & 1 \\ 1 & 1 \end{bmatrix}$. Then $B^{-1}A = \begin{bmatrix} -2 & 1 \\ 5 & 1 \end{bmatrix}$.

Nevertheless, the eigenvalues of $B^{-1}A$ are always real (and are approximately 2.1926 and -3.1926 in Example 12.16).

Theorem 12.17. *Let $A, B \in \mathbb{R}^{n \times n}$ with $A = A^T$ and $B = B^T > 0$. Then the generalized eigenvalue problem*

$$Ax = \lambda Bx$$

has n real eigenvalues, and the n corresponding right eigenvectors can be chosen to be orthogonal with respect to the inner product $\langle x, y \rangle_B = x^T By$. Moreover, if $A > 0$, then the eigenvalues are also all positive.

Proof: Since $B > 0$, it has a Cholesky factorization $B = LL^T$, where L is nonsingular (Theorem 10.23). Then the eigenvalue problem

$$Ax = \lambda Bx = \lambda LL^T x$$

can be rewritten as the equivalent problem

$$(L^{-1}AL^{-T})(L^T x) = \lambda L^T x. \tag{12.11}$$

Letting $C = L^{-1}AL^{-T}$ and $z = L^T x$, (12.11) can then be rewritten as

$$Cz = \lambda z. \tag{12.12}$$

Since $C = C^T$, the eigenproblem (12.12) has n real eigenvalues, with corresponding eigenvectors $z_1, \ldots, z_n$ satisfying

$$z_i^T z_j = \delta_{ij}.$$

Then $x_i = L^{-T} z_i$, $i \in \underline{n}$, are eigenvectors of the original generalized eigenvalue problem and satisfy

$$\langle x_i, x_j \rangle_B = x_i^T B x_j = (z_i^T L^{-1})(LL^T)(L^{-T} z_j) = \delta_{ij}.$$

Finally, if $A = A^T > 0$, then $C = C^T > 0$, so the eigenvalues are positive. $\square$

Example 12.18. The Cholesky factor for the matrix B in Example 12.16 is

$$L = \begin{bmatrix} \sqrt{2} & 0 \\ \frac{1}{\sqrt{2}} & \frac{1}{\sqrt{2}} \end{bmatrix}.$$

Then it is easily checked that

$$C = L^{-1}AL^{-T} = \begin{bmatrix} 0.5 & 2.5 \\ 2.5 & -1.5 \end{bmatrix},$$

whose eigenvalues are approximately 2.1926 and -3.1926 as expected.

The material of this section can, of course, be generalized easily to the case where A and B are Hermitian, but since real-valued matrices are commonly used in most applications, we have restricted our attention to that case only.

12.5 Simultaneous Diagonalization

Recall that many matrices can be diagonalized by a similarity. In particular, normal matrices can be diagonalized by a unitary similarity. It turns out that in some cases a pair of matrices (A, B) can be simultaneously diagonalized by the same matrix. There are many such results and we present only a representative (but important and useful) theorem here. Again, we restrict our attention only to the real case, with the complex case following in a straightforward way.

Theorem 12.19 (Simultaneous Reduction to Diagonal Form). *Let $A, B \in \mathbb{R}^{n\times n}$ with $A = A^T$ and $B = B^T > 0$. Then there exists a nonsingular matrix Q such that*

$$Q^T AQ = D, \quad Q^T BQ = I,$$

where D is diagonal. In fact, the diagonal elements of D are the eigenvalues of $B^{-1}A$.

Proof: Let $B = LL^T$ be the Cholesky factorization of B and set $C = L^{-1}AL^{-T}$. Since C is symmetric, there exists an orthogonal matrix P such that $P^TCP = D$, where D is diagonal. Let $Q = L^{-T}P$. Then

$$Q^T AQ = P^T L^{-1} AL^{-T} P = P^T CP = D$$

and

$$Q^T BQ = P^T L^{-1}(LL^T)L^{-T} P = P^T P = I.$$

Finally, since $QDQ^{-1} = QQ^TAQQ^{-1} = L^{-T}PP^TL^{-1}A = L^{-T}L^{-1}A = B^{-1}A$, we have $\Lambda(D) = \Lambda(B^{-1}A)$. □

Note that Q is not in general orthogonal, so it does not preserve eigenvalues of A and B individually. However, it does preserve the eigenvalues of $A - \lambda B$. This can be seen directly. Let $\tilde{A} = Q^TAQ$ and $\tilde{B} = Q^TBQ$. Then $\tilde{B}^{-1}\tilde{A} = Q^{-1}B^{-1}Q^{-T}Q^TAQ = Q^{-1}B^{-1}AQ$.

Theorem 12.19 is very useful for reducing many statements about pairs of symmetric matrices to "the diagonal case." The following is typical.

Theorem 12.20. *Let $A, B \in \mathbb{R}^{n\times n}$ be positive definite. Then $A \geq B$ if and only if $B^{-1} \geq A^{-1}$.*

Proof: By Theorem 12.19, there exists $Q \in \mathbb{R}_n^{n\times n}$ such that $Q^TAQ = D$ and $Q^TBQ = I$, where D is diagonal. Now $D > 0$ by Theorem 10.31. Also, since $A \geq B$, by Theorem 10.21 we have that $Q^TAQ \geq Q^TBQ$, i.e., $D \geq I$. But then $D^{-1} \leq I$ (this is trivially true since the two matrices are diagonal). Thus, $QD^{-1}Q^T \leq QQ^T$, i.e., $A^{-1} \leq B^{-1}$. □

12.5.1 Simultaneous diagonalization via SVD

There are situations in which forming $C = L^{-1}AL^{-T}$ as in the proof of Theorem 12.19 is numerically problematic, e.g., when L is highly ill conditioned with respect to inversion. In such cases, simultaneous reduction can also be accomplished via an SVD. To illustrate, let us assume that both A and B are positive definite. Further, let $A = L_A L_A^T$ and $B = L_B L_B^T$ be Cholesky factorizations of A and B, respectively. Compute the SVD

$$L_B^{-1} L_A = U \Sigma V^T, \tag{12.13}$$

where $\Sigma \in \mathbb{R}_n^{n \times n}$ is diagonal. Then the matrix $Q = L_B^{-T} U$ performs the simultaneous diagonalization. To check this, note that

$$\begin{aligned} Q^T A Q &= U^T L_B^{-1} (L_A L_A^T) L_B^{-T} U \\ &= U^T U \Sigma V^T V \Sigma^T U^T U \\ &= \Sigma^2 \end{aligned}$$

while

$$\begin{aligned} Q^T B Q &= U^T L_B^{-1} (L_B L_B^T) L_B^{-T} U \\ &= U^T U \\ &= I. \end{aligned}$$

Remark 12.21. The SVD in (12.13) can be computed without explicitly forming the indicated matrix product or the inverse by using the so-called **generalized singular value decomposition** (GSVD). Note that

$$\sigma(L_B^{-1} L_A) = \lambda^{\frac{1}{2}} (L_B^{-1} L_A L_A^T L_B^{-T})$$

and thus the singular values of $L_B^{-1} L_A$ can be found from the eigenvalue problem

$$L_B^{-1} L_A L_A^T L_B^{-T} z = \lambda z. \tag{12.14}$$

Letting $x = L_B^{-T} z$ we see that (12.14) can be rewritten in the form $L_A L_A^T x = \lambda L_B z = \lambda L_B L_B^T L_B^{-T} z$, which is thus equivalent to the generalized eigenvalue problem

$$L_A L_A^T x = \lambda L_B L_B^T x. \tag{12.15}$$

The problem (12.15) is called a generalized singular value problem and algorithms exist to solve it (and hence equivalently (12.13)) via arithmetic operations performed only on L_A and L_B separately, i.e., without forming the products $L_A L_A^T$ or $L_B L_B^T$ explicitly; see, for example, [7, Sec. 8.7.3]. This is analogous to finding the singular values of a matrix M by operations performed directly on M rather than by forming the matrix $M^T M$ and solving the eigenproblem $M^T M x = \lambda x$.

Remark 12.22. Various generalizations of the results in Remark 12.21 are possible, for example, when $A = A^T \geq 0$. The case when A is symmetric but indefinite is not so straightforward, at least in real arithmetic. For example, A can be written as $A = P D P^T$, where D is diagonal and P is orthogonal, but in writing $A = P \tilde{D} \tilde{D} P^T = P \tilde{D} (P \tilde{D})^T$ with $\tilde{D}$ diagonal, $\tilde{D}$ may have pure imaginary elements.

12.6 Higher-Order Eigenvalue Problems

Consider the second-order system of differential equations

$$M\ddot{q} + C\dot{q} + Kq = 0, \tag{12.16}$$

where $q(t) \in \mathbb{R}^n$ and $M, C, K \in \mathbb{R}^{n \times n}$. Assume for simplicity that M is nonsingular. Suppose, by analogy with the first-order case, that we try to find a solution of (12.16) of the form $q(t) = e^{\lambda t} p$, where the n-vector p and scalar λ are to be determined. Substituting in (12.16) we get

$$\lambda^2 e^{\lambda t} Mp + \lambda e^{\lambda t} Cp + e^{\lambda t} Kp = 0$$

or, since $e^{\lambda t} \neq 0$,

$$(\lambda^2 M + \lambda C + K)p = 0.$$

To get a nonzero solution p, we thus seek values of λ for which the matrix $\lambda^2 M + \lambda C + K$ is singular. Since the determinantal equation

$$0 = \det(\lambda^2 M + \lambda C + K) = \lambda^{2n} + \cdots$$

yields a polynomial of degree $2n$, there are $2n$ eigenvalues for the **second-order** (or **quadratic) eigenvalue problem** $\lambda^2 M + \lambda C + K$.

A special case of (12.16) arises frequently in applications: $M = I$, $C = 0$, and $K = K^T$. Suppose K has eigenvalues

$$\mu_1 \geq \cdots \geq \mu_r \geq 0 > \mu_{r+1} \geq \cdots \geq \mu_n.$$

Let $\omega_k = |\mu_k|^{\frac{1}{2}}$. Then the $2n$ eigenvalues of the second-order eigenvalue problem $\lambda^2 I + K$ are

$$\begin{aligned} &\pm\, j\omega_k;\, k = 1, \ldots, r, \\ &\pm\, \omega_k;\, k = r+1, \ldots, n. \end{aligned}$$

If $r = n$ (i.e., $K = K^T \geq 0$), then all solutions of $\ddot{q} + Kq = 0$ are oscillatory.

12.6.1 Conversion to first-order form

Let $x_1 = q$ and $x_2 = \dot{q}$. Then (12.16) can be written as a first-order system (with block companion matrix)

$$\dot{x} = \begin{bmatrix} 0 & I \\ -M^{-1}K & -M^{-1}C \end{bmatrix} x,$$

where $x(t) \in \mathbb{R}^{2n}$. If M is singular, or if it is desired to avoid the calculation of M^{-1} because M is too ill conditioned with respect to inversion, the second-order problem (12.16) can still be converted to the first-order generalized linear system

$$\begin{bmatrix} I & 0 \\ 0 & M \end{bmatrix} \dot{x} = \begin{bmatrix} 0 & I \\ -K & -C \end{bmatrix} x.$$

Many other first-order realizations are possible. Some can be useful when M, C, and/or K have special symmetry or skew-symmetry properties that can exploited.

Higher-order analogues of (12.16) involving, say, the kth derivative of q, lead naturally to higher-order eigenvalue problems that can be converted to first-order form using a $kn \times kn$ block companion matrix analogue of (11.19). Similar procedures hold for the general kth-order difference equation

$$A_k q^{(k)}(t) + A_{k-1} q^{(k-1)}(t) + \cdots + A_1 \dot{q}(t) + A_0 q(t) = 0,$$

which can be converted to various first-order systems of dimension kn.

EXERCISES

1. Suppose $A \in \mathbb{R}^{n\times n}$ and $D \in \mathbb{R}_m^{m\times m}$. Show that the finite generalized eigenvalues of the pencil
$$\begin{bmatrix} A & B \\ C & D \end{bmatrix} - \lambda \begin{bmatrix} I & 0 \\ 0 & 0 \end{bmatrix}$$
are the eigenvalues of the matrix $A - BD^{-1}C$.

2. Let $F, G \in \mathbb{C}^{n\times n}$. Show that the nonzero eigenvalues of FG and GF are the same. *Hint*: An easy "trick proof" is to verify that the matrices
$$\begin{bmatrix} FG & 0 \\ G & 0 \end{bmatrix} \text{ and } \begin{bmatrix} 0 & 0 \\ G & GF \end{bmatrix}$$
are similar via the similarity transformation
$$\begin{bmatrix} I & F \\ 0 & I \end{bmatrix}.$$

3. Let $F \in \mathbb{C}^{n\times m}$, $G \in \mathbb{C}^{m\times n}$. Are the nonzero singular values of FG and GF the same?

4. Suppose $A \in \mathbb{R}^{n\times n}$, $B \in \mathbb{R}^{n\times m}$, and $C \in \mathbb{R}^{m\times n}$. Show that the generalized eigenvalues of the pencils
$$\begin{bmatrix} A & B \\ C & 0 \end{bmatrix} - \lambda \begin{bmatrix} I & 0 \\ 0 & 0 \end{bmatrix}$$
and
$$\begin{bmatrix} A + BF + GC & B \\ C & 0 \end{bmatrix} - \lambda \begin{bmatrix} I & 0 \\ 0 & 0 \end{bmatrix}$$
are identical for all $F \in \mathbb{R}^{m\times n}$ and all $G \in \mathbb{R}^{n\times m}$.
Hint: Consider the equivalence
$$\begin{bmatrix} I & G \\ 0 & I \end{bmatrix} \begin{bmatrix} A - \lambda I & B \\ C & 0 \end{bmatrix} \begin{bmatrix} I & 0 \\ F & I \end{bmatrix}.$$
(A similar result is also true for "nonsquare" pencils. In the parlance of control theory, such results show that zeros are invariant under state feedback or output injection.)

5. Another family of simultaneous diagonalization problems arises when it is desired that the simultaneous diagonalizing transformation Q operates on matrices $A, B \in \mathbb{R}^{n\times n}$ in such a way that $Q^{-1}AQ^{-T}$ and Q^TBQ are simultaneously diagonal. Such a transformation is called **contragredient**. Consider the case where both A and B are positive definite with Cholesky factorizations $A = L_A L_A^T$ and $B = L_B L_B^T$, respectively, and let $U\Sigma V^T$ be an SVD of $L_B^T L_A$.

   (a) Show that $Q = L_A V \Sigma^{-\frac{1}{2}}$ is a contragredient transformation that reduces both A and B to the same diagonal matrix.

   (b) Show that $Q^{-1} = \Sigma^{-\frac{1}{2}} U^T L_B^T$.

   (c) Show that the eigenvalues of AB are the same as those of Σ^2 and hence are positive.

Chapter 13

Kronecker Products

13.1 Definition and Examples

Definition 13.1. *Let $A \in \mathbb{R}^{m \times n}$, $B \in \mathbb{R}^{p \times q}$. Then the* **Kronecker product** *(or tensor product) of A and B is defined as the matrix*

$$A \otimes B = \begin{bmatrix} a_{11}B & \cdots & a_{1n}B \\ \vdots & \ddots & \vdots \\ a_{m1}B & \cdots & a_{mn}B \end{bmatrix} \in \mathbb{R}^{mp \times nq}. \tag{13.1}$$

Obviously, the same definition holds if A and B are complex-valued matrices. We restrict our attention in this chapter primarily to real-valued matrices, pointing out the extension to the complex case only where it is not obvious.

Example 13.2.

1. Let $A = \begin{bmatrix} 1 & 2 & 3 \\ 3 & 2 & 1 \end{bmatrix}$ and $B = \begin{bmatrix} 2 & 1 \\ 2 & 3 \end{bmatrix}$. Then

$$A \otimes B = \begin{bmatrix} B & 2B & 3B \\ 3B & 2B & B \end{bmatrix} = \begin{bmatrix} 2 & 1 & 4 & 2 & 6 & 3 \\ 2 & 3 & 4 & 6 & 6 & 9 \\ 6 & 3 & 4 & 2 & 2 & 1 \\ 6 & 9 & 4 & 6 & 2 & 3 \end{bmatrix}.$$

   Note that $B \otimes A \neq A \otimes B$.

2. For any $B \in \mathbb{R}^{p \times q}$, $I_2 \otimes B = \begin{bmatrix} B & 0 \\ 0 & B \end{bmatrix}$.
   Replacing I_2 by I_n yields a block diagonal matrix with n copies of B along the diagonal.

3. Let B be an arbitrary 2×2 matrix. Then

$$B \otimes I_2 = \begin{bmatrix} b_{11} & 0 & b_{12} & 0 \\ 0 & b_{11} & 0 & b_{12} \\ b_{21} & 0 & b_{22} & 0 \\ 0 & b_{21} & 0 & b_{22} \end{bmatrix}.$$

The extension to arbitrary B and I_n is obvious.

4. Let $x \in \mathbb{R}^m$, $y \in \mathbb{R}^n$. Then

$$\begin{aligned} x \otimes y &= \left[x_1 y^T, \dots, x_m y^T\right]^T \\ &= [x_1 y_1, \dots, x_1 y_n, x_2 y_1, \dots, x_m y_n]^T \in \mathbb{R}^{mn}. \end{aligned}$$

5. Let $x \in \mathbb{R}^m$, $y \in \mathbb{R}^n$. Then

$$\begin{aligned} x \otimes y^T &= [x_1 y, \dots, x_m y]^T \\ &= \begin{bmatrix} x_1 y_1 & \dots & x_1 y_n \\ \vdots & \ddots & \vdots \\ x_m y_1 & \dots & x_m y_n \end{bmatrix} \\ &= x y^T \in \mathbb{R}^{m \times n}. \end{aligned}$$

13.2 Properties of the Kronecker Product

Theorem 13.3. *Let $A \in \mathbb{R}^{m \times n}$, $B \in \mathbb{R}^{r \times s}$, $C \in \mathbb{R}^{n \times p}$, and $D \in \mathbb{R}^{s \times t}$. Then*

$$(A \otimes B)(C \otimes D) = AC \otimes BD \quad (\in \mathbb{R}^{mr \times pt}). \tag{13.2}$$

Proof: Simply verify that

$$\begin{aligned} (A \otimes B)(C \otimes D) &= \begin{bmatrix} a_{11}B & \cdots & a_{1n}B \\ \vdots & \ddots & \vdots \\ a_{m1}B & \cdots & a_{mn}B \end{bmatrix} \begin{bmatrix} c_{11}D & \cdots & c_{1p}D \\ \vdots & \ddots & \vdots \\ c_{n1}D & \cdots & c_{np}D \end{bmatrix} \\ &= \begin{bmatrix} \sum_{k=1}^n a_{1k}c_{k1}BD & \cdots & \sum_{k=1}^n a_{1k}c_{kp}BD \\ \vdots & \ddots & \vdots \\ \sum_{k=1}^n a_{mk}c_{k1}BD & \cdots & \sum_{k=1}^n a_{mk}c_{kp}BD \end{bmatrix} \\ &= AC \otimes BD. \qquad \square \end{aligned}$$

Theorem 13.4. *For all A and B, $(A \otimes B)^T = A^T \otimes B^T$.*

Proof: For the proof, simply verify using the definitions of transpose and Kronecker product. $\square$

Corollary 13.5. *If $A \in \mathbb{R}^{n \times n}$ and $B \in \mathbb{R}^{m \times m}$ are symmetric, then $A \otimes B$ is symmetric.*

Theorem 13.6. *If A and B are nonsingular, $(A \otimes B)^{-1} = A^{-1} \otimes B^{-1}$.*

Proof: Using Theorem 13.3, simply note that $(A \otimes B)(A^{-1} \otimes B^{-1}) = I \otimes I = I$. $\square$

Theorem 13.7. *If $A \in \mathbb{R}^{n\times n}$ and $B \in \mathbb{R}^{m\times m}$ are normal, then $A \otimes B$ is normal.*

Proof:

$$\begin{aligned}(A\otimes B)^T(A\otimes B) &= (A^T \otimes B^T)(A\otimes B) \quad \text{by Theorem 13.4}\\ &= A^TA \otimes B^TB \quad \text{by Theorem 13.3}\\ &= AA^T \otimes BB^T \quad \text{since } A \text{ and } B \text{ are normal}\\ &= (A\otimes B)(A\otimes B)^T \quad \text{by Theorem 13.3.} \qquad \square\end{aligned}$$

Corollary 13.8. *If $A \in \mathbb{R}^{n\times n}$ is orthogonal and $B \in \mathbb{R}^{m\times m}$ is orthogonal, then $A \otimes B$ is orthogonal.*

Example 13.9. Let $A = \begin{bmatrix} \cos\theta & \sin\theta \\ -\sin\theta & \cos\theta \end{bmatrix}$ and $B = \begin{bmatrix} \cos\phi & \sin\phi \\ -\sin\phi & \cos\phi \end{bmatrix}$. Then it is easily seen that A is orthogonal with eigenvalues $e^{\pm j\theta}$ and B is orthogonal with eigenvalues $e^{\pm j\phi}$. The 4×4 matrix $A\otimes B$ is then also orthogonal with eigenvalues $e^{\pm j(\theta+\phi)}$ and $e^{\pm j(\theta-\phi)}$.

Theorem 13.10. *Let $A \in \mathbb{R}^{m\times n}$ have a singular value decomposition $U_A\Sigma_AV_A^T$ and let $B \in \mathbb{R}^{p\times q}$ have a singular value decomposition $U_B\Sigma_BV_B^T$. Then*

$$(U_A \otimes U_B)(\Sigma_A \otimes \Sigma_B)(V_A^T \otimes V_B^T)$$

yields a singular value decomposition of $A \otimes B$ (after a simple reordering of the diagonal elements of $\Sigma_A \otimes \Sigma_B$ and the corresponding right and left singular vectors).

Corollary 13.11. *Let $A \in \mathbb{R}_r^{m\times n}$ have singular values $\sigma_1 \geq \cdots \geq \sigma_r > 0$ and let $B \in \mathbb{R}_s^{p\times q}$ have singular values $\tau_1 \geq \cdots \geq \tau_s > 0$. Then $A\otimes B$ (or $B \otimes A$) has rs singular values $\sigma_1\tau_1 \geq \cdots \geq \sigma_r\tau_s > 0$ and*

$$\operatorname{rank}(A\otimes B) = (\operatorname{rank} A)(\operatorname{rank} B) = \operatorname{rank}(B\otimes A)\,.$$

Theorem 13.12. *Let $A \in \mathbb{R}^{n\times n}$ have eigenvalues λ_i, $i \in \underline{n}$, and let $B \in \mathbb{R}^{m\times m}$ have eigenvalues μ_j, $j \in \underline{m}$. Then the mn eigenvalues of $A\otimes B$ are*

$$\lambda_1\mu_1, \ldots, \lambda_1\mu_m, \lambda_2\mu_1, \ldots, \lambda_2\mu_m, \ldots, \lambda_n\mu_m.$$

Moreover, if $x_1, \ldots, x_p$ are linearly independent right eigenvectors of A corresponding to $\lambda_1, \ldots, \lambda_p$ ($p \leq n$), and $z_1, \ldots, z_q$ are linearly independent right eigenvectors of B corresponding to $\mu_1, \ldots, \mu_q$ ($q \leq m$), then $x_i \otimes z_j \in \mathbb{R}^{mn}$ are linearly independent right eigenvectors of $A\otimes B$ corresponding to $\lambda_i\mu_j$, $i \in \underline{p}$, $j \in \underline{q}$.

Proof: The basic idea of the proof is as follows:

$$\begin{aligned}(A\otimes B)(x\otimes z) &= Ax \otimes Bz\\ &= \lambda x \otimes \mu z\\ &= \lambda\mu(x\otimes z). \qquad \square\end{aligned}$$

If A and B are diagonalizable in Theorem 13.12, we can take $p = n$ and $q = m$ and thus get the complete eigenstructure of $A \otimes B$. In general, if A and B have Jordan form decompositions given by $P^{-1}AP = J_A$ and $Q^{-1}BQ = J_B$, respectively, then we get the following Jordan-like structure:

$$\begin{aligned}(P \otimes Q)^{-1}(A \otimes B)(P \otimes Q) &= (P^{-1} \otimes Q^{-1})(A \otimes B)(P \otimes Q) \\ &= (P^{-1}AP) \otimes (Q^{-1}BQ) \\ &= J_A \otimes J_B.\end{aligned}$$

Note that $J_A \otimes J_B$, while upper triangular, is generally not quite in Jordan form and needs further reduction (to an ultimate Jordan form that also depends on whether or not certain eigenvalues are zero or nonzero).

A Schur form for $A \otimes B$ can be derived similarly. For example, suppose P and Q are unitary matrices that reduce A and B, respectively, to Schur (triangular) form, i.e., $P^H AP = T_A$ and $Q^H BQ = T_B$ (and similarly if P and Q are orthogonal similarities reducing A and B to real Schur form). Then

$$\begin{aligned}(P \otimes Q)^{H}(A \otimes B)(P \otimes Q) &= (P^{H} \otimes Q^{H})(A \otimes B)(P \otimes Q) \\ &= (P^{H}AP) \otimes (Q^{H}BQ) \\ &= T_A \otimes T_B.\end{aligned}$$

Corollary 13.13. *Let $A \in \mathbb{R}^{n \times n}$ and $B \in \mathbb{R}^{m \times m}$. Then*

1. $\operatorname{Tr}(A \otimes B) = (\operatorname{Tr} A)(\operatorname{Tr} B) = \operatorname{Tr}(B \otimes A)$.

2. $\det(A \otimes B) = (\det A)^m (\det B)^n = \det(B \otimes A)$.

Definition 13.14. *Let $A \in \mathbb{R}^{n \times n}$ and $B \in \mathbb{R}^{m \times m}$. Then the* **Kronecker sum** *(or tensor sum) of A and B, denoted $A \oplus B$, is the $mn \times mn$ matrix $(I_m \otimes A) + (B \otimes I_n)$. Note that, in general, $A \oplus B \neq B \oplus A$.*

Example 13.15.

1. Let

$$A = \begin{bmatrix} 1 & 2 & 3 \\ 3 & 2 & 1 \\ 1 & 1 & 4 \end{bmatrix} \text{ and } B = \begin{bmatrix} 2 & 1 \\ 2 & 3 \end{bmatrix}.$$

Then

$$A \oplus B = (I_2 \otimes A) + (B \otimes I_3) = \begin{bmatrix} 1 & 2 & 3 & 0 & 0 & 0 \\ 3 & 2 & 1 & 0 & 0 & 0 \\ 1 & 1 & 4 & 0 & 0 & 0 \\ 0 & 0 & 0 & 1 & 2 & 3 \\ 0 & 0 & 0 & 3 & 2 & 1 \\ 0 & 0 & 0 & 1 & 1 & 4 \end{bmatrix} + \begin{bmatrix} 2 & 0 & 0 & 1 & 0 & 0 \\ 0 & 2 & 0 & 0 & 1 & 0 \\ 0 & 0 & 2 & 0 & 0 & 1 \\ 2 & 0 & 0 & 3 & 0 & 0 \\ 0 & 2 & 0 & 0 & 3 & 0 \\ 0 & 0 & 2 & 0 & 0 & 3 \end{bmatrix}.$$

The reader is invited to compute $B \oplus A = (I_3 \otimes B) + (A \otimes I_2)$ and note the difference with $A \oplus B$.

2. Recall the real JCF

$$J = \begin{bmatrix} M & I & 0 & \cdots & & 0 \\ 0 & M & I & 0 & & \vdots \\ \vdots & \ddots & M & \ddots & \ddots & \\ & & & \ddots & I & 0 \\ \vdots & & & \ddots & M & I \\ 0 & \cdots & & \cdots & 0 & M \end{bmatrix} \in \mathbb{R}^{2k \times 2k},$$

where $M = \begin{bmatrix} \alpha & \beta \\ -\beta & \alpha \end{bmatrix}$. Define

$$E_k = \begin{bmatrix} 0 & 1 & 0 & \cdots & 0 \\ 0 & 0 & 1 & \ddots & \vdots \\ \vdots & & \ddots & \ddots & 0 \\ & & & \ddots & 1 \\ 0 & \cdots & & \cdots & 0 \end{bmatrix} \in \mathbb{R}^{k \times k}.$$

Then J can be written in the very compact form $J = (I_k \otimes M) + (E_k \otimes I_2) = M \oplus E_k$.

Theorem 13.16. *Let $A \in \mathbb{R}^{n \times n}$ have eigenvalues $\lambda_i, i \in \underline{n}$, and let $B \in \mathbb{R}^{m \times m}$ have eigenvalues $\mu_j, j \in \underline{m}$. Then the Kronecker sum $A \oplus B = (I_m \otimes A) + (B \otimes I_n)$ has mn eigenvalues*

$$\lambda_1 + \mu_1, \ldots, \lambda_1 + \mu_m, \lambda_2 + \mu_1, \ldots, \lambda_2 + \mu_m, \ldots, \lambda_n + \mu_m.$$

Moreover, if $x_1, \ldots, x_p$ are linearly independent right eigenvectors of A corresponding to $\lambda_1, \ldots, \lambda_p$ $(p \le n)$, and $z_1, \ldots, z_q$ are linearly independent right eigenvectors of B corresponding to $\mu_1, \ldots, \mu_q$ $(q \le m)$, then $z_j \otimes x_i \in \mathbb{R}^{mn}$ are linearly independent right eigenvectors of $A \oplus B$ corresponding to $\lambda_i + \mu_j$, $i \in \underline{p}$, $j \in \underline{q}$.

Proof: The basic idea of the proof is as follows:

$$\begin{aligned} [(I_m \otimes A) + (B \otimes I_n)](z \otimes x) &= (z \otimes Ax) + (Bz \otimes x) \\ &= (z \otimes \lambda x) + (\mu z \otimes x) \\ &= (\lambda + \mu)(z \otimes x). \qquad \square \end{aligned}$$

If A and B are diagonalizable in Theorem 13.16, we can take $p = n$ and $q = m$ and thus get the complete eigenstructure of $A \oplus B$. In general, if A and B have Jordan form decompositions given by $P^{-1}AP = J_A$ and $Q^{-1}BQ = J_B$, respectively, then

$$\begin{aligned} &[(Q \otimes I_n)(I_m \otimes P)]^{-1}[(I_m \otimes A) + (B \otimes I_n)][(Q \otimes I_n)(I_m \otimes P)] \\ &= [(I_m \otimes P)^{-1}(Q \otimes I_n)^{-1}][(I_m \otimes A) + (B \otimes I_n)][(Q \otimes I_n)(I_m \otimes P)] \\ &= [(I_m \otimes P^{-1})(Q^{-1} \otimes I_n)][(I_m \otimes A) + (B \otimes I_n)][(Q \otimes I_n)(I_m \otimes P)] \\ &= (I_m \otimes J_A) + (J_B \otimes I_n) \end{aligned}$$

is a Jordan-like structure for $A \oplus B$.

A Schur form for $A \oplus B$ can be derived similarly. Again, suppose P and Q are unitary matrices that reduce A and B, respectively, to Schur (triangular) form, i.e., $P^H AP = T_A$ and $Q^H BQ = T_B$ (and similarly if P and Q are orthogonal similarities reducing A and B to real Schur form). Then

$$[(Q \otimes I_n)(I_m \otimes P)]^H [(I_m \otimes A) + (B \otimes I_n)][(Q \otimes I_n)(I_m \otimes P)] = (I_m \otimes T_A) + (T_B \otimes I_n),$$

where $[(Q \otimes I_n)(I_m \otimes P)] = (Q \otimes P)$ is unitary by Theorem 13.3 and Corollary 13.8.

13.3 Application to Sylvester and Lyapunov Equations

In this section we study the linear matrix equation

$$AX + XB = C, \tag{13.3}$$

where $A \in \mathbb{R}^{n\times n}$, $B \in \mathbb{R}^{m\times m}$, and $C \in \mathbb{R}^{n\times m}$. This equation is now often called a **Sylvester equation** in honor of J.J. Sylvester who studied general linear matrix equations of the form

$$\sum_{i=1}^{k} A_i X B_i = C.$$

A special case of (13.3) is the symmetric equation

$$AX + XA^T = C \tag{13.4}$$

obtained by taking $B = A^T$. When C is symmetric, the solution $X \in \mathbb{R}^{n\times n}$ is easily shown also to be symmetric and (13.4) is known as a **Lyapunov equation**. Lyapunov equations arise naturally in stability theory.

The first important question to ask regarding (13.3) is, When does a solution exist? By writing the matrices in (13.3) in terms of their columns, it is easily seen by equating the ith columns that

$$Ax_i + Xb_i = c_i = Ax_i + \sum_{j=1}^{m} b_{ji} x_j.$$

These equations can then be rewritten as the $mn \times mn$ linear system

$$\begin{bmatrix} A + b_{11}I & b_{21}I & \cdots & b_{m1}I \\ b_{12}I & A + b_{22}I & \cdots & b_{m2}I \\ \vdots & & \ddots & \vdots \\ b_{1m}I & b_{2m}I & \cdots & A + b_{mm}I \end{bmatrix} \begin{bmatrix} x_1 \\ \vdots \\ x_m \end{bmatrix} = \begin{bmatrix} c_1 \\ \vdots \\ c_m \end{bmatrix}. \tag{13.5}$$

The coefficient matrix in (13.5) clearly can be written as the Kronecker sum $(I_m \otimes A) + (B^T \otimes I_n)$. The following definition is very helpful in completing the writing of (13.5) as an "ordinary" linear system.

Definition 13.17. *Let $c_i \in \mathbb{R}^n$ denote the columns of $C \in \mathbb{R}^{n \times m}$ so that $C = [c_1, \ldots, c_m]$. Then* $\text{vec}(C)$ *is defined to be the mn-vector formed by stacking the columns of C on top of one another, i.e.,* $\text{vec}(C) = \begin{bmatrix} c_1 \\ \vdots \\ c_m \end{bmatrix} \in \mathbb{R}^{mn}$.

Using Definition 13.17, the linear system (13.5) can be rewritten in the form

$$[(I_m \otimes A) + (B^T \otimes I_n)]\text{vec}(X) = \text{vec}(C). \tag{13.6}$$

There exists a unique solution to (13.6) if and only if $[(I_m \otimes A) + (B^T \otimes I_n)]$ is nonsingular. But $[(I_m \otimes A) + (B^T \otimes I_n)]$ is nonsingular if and only if it has no zero eigenvalues. From Theorem 13.16, the eigenvalues of $[(I_m \otimes A) + (B^T \otimes I_n)]$ are $\lambda_i + \mu_j$, where $\lambda_i \in \Lambda(A)$, $i \in \underline{n}$, and $\mu_j \in \Lambda(B)$, $j \in \underline{m}$. We thus have the following theorem.

Theorem 13.18. *Let $A \in \mathbb{R}^{n \times n}$, $B \in \mathbb{R}^{m \times m}$, and $C \in \mathbb{R}^{n \times m}$. Then the Sylvester equation*

$$AX + XB = C \tag{13.7}$$

has a unique solution if and only if A and $-B$ have no eigenvalues in common.

Sylvester equations of the form (13.3) (or symmetric Lyapunov equations of the form (13.4)) are generally not solved using the $mn \times mn$ "vec" formulation (13.6). The most commonly preferred numerical algorithm is described in [2]. First A and B are reduced to (real) Schur form. An equivalent linear system is then solved in which the triangular form of the reduced A and B can be exploited to solve successively for the columns of a suitably transformed solution matrix X. Assuming that, say, $n \geq m$, this algorithm takes only $O(n^3)$ operations rather than the $O(n^6)$ that would be required by solving (13.6) directly with Gaussian elimination. A further enhancement to this algorithm is available in [6] whereby the larger of A or B is initially reduced only to upper Hessenberg rather than triangular Schur form.

The next few theorems are classical. They culminate in Theorem 13.24, one of many elegant connections between matrix theory and stability theory for differential equations.

Theorem 13.19. *Let $A \in \mathbb{R}^{n \times n}$, $B \in \mathbb{R}^{m \times m}$, and $C \in \mathbb{R}^{n \times m}$. Suppose further that A and B are* **asymptotically stable** *(a matrix is asymptotically stable if all its eigenvalues have real parts in the open left half-plane). Then the (unique) solution of the Sylvester equation*

$$AX + XB = C \tag{13.8}$$

can be written as

$$X = -\int_0^{+\infty} e^{tA} C e^{tB}\, dt. \tag{13.9}$$

Proof: Since A and B are stable, $\lambda_i(A) + \lambda_j(B) \neq 0$ for all i, j so there exists a unique solution to (13.8) by Theorem 13.18. Now integrate the differential equation $\dot{X} = AX + XB$ (with $X(0) = C$) on $[0, +\infty)$:

$$\lim_{t \to +\infty} X(t) - X(0) = A \int_0^{+\infty} X(t)\, dt + \left(\int_0^{+\infty} X(t)\, dt \right) B. \tag{13.10}$$

Using the results of Section 11.1.6, it can be shown easily that $\lim_{t \to +\infty} e^{tA} = \lim_{t \to +\infty} e^{tB} = 0$. Hence, using the solution $X(t) = e^{tA}Ce^{tB}$ from Theorem 11.6, we have that $\lim_{t \to +\infty} X(t) = 0$. Substituting in (13.10) we have

$$-C = A\left(\int_0^{+\infty} e^{tA}Ce^{tB}\,dt\right) + \left(\int_0^{+\infty} e^{tA}Ce^{tB}\,dt\right)B$$

and so $X = -\int_0^{+\infty} e^{tA}Ce^{tB}\,dt$ satisfies (13.8). □

Remark 13.20. An equivalent condition for the existence of a unique solution to $AX + XB = C$ is that $\begin{bmatrix} A & C \\ 0 & -B \end{bmatrix}$ be similar to $\begin{bmatrix} A & 0 \\ 0 & -B \end{bmatrix}$ (via the similarity $\begin{bmatrix} I & X \\ 0 & -I \end{bmatrix}$).

Theorem 13.21. *Let $A, C \in \mathbb{R}^{n\times n}$. Then the Lyapunov equation*

$$AX + XA^T = C \tag{13.11}$$

has a unique solution if and only if A and $-A^T$ have no eigenvalues in common. If C is symmetric and (13.11) has a unique solution, then that solution is symmetric.

Remark 13.22. If the matrix $A \in \mathbb{R}^{n\times n}$ has eigenvalues $\lambda_1, \ldots, \lambda_n$, then $-A^T$ has eigenvalues $-\lambda_1, \ldots, -\lambda_n$. Thus, a sufficient condition that guarantees that A and $-A^T$ have no common eigenvalues is that A be asymptotically stable. Many useful results exist concerning the relationship between stability and Lyapunov equations. Two basic results due to Lyapunov are the following, the first of which follows immediately from Theorem 13.19.

Theorem 13.23. *Let $A, C \in \mathbb{R}^{n\times n}$ and suppose further that A is asymptotically stable. Then the (unique) solution of the Lyapunov equation*

$$AX + XA^T = C$$

can be written as

$$X = -\int_0^{+\infty} e^{tA}Ce^{tA^T}\,dt. \tag{13.12}$$

Theorem 13.24. *A matrix $A \in \mathbb{R}^{n\times n}$ is asymptotically stable if and only if there exists a positive definite solution to the Lyapunov equation*

$$AX + XA^T = C, \tag{13.13}$$

where $C = C^T < 0$.

Proof: Suppose A is asymptotically stable. By Theorems 13.21 and 13.23 a solution to (13.13) exists and takes the form (13.12). Now let v be an arbitrary nonzero vector in $\mathbb{R}^n$. Then

$$v^T Xv = \int_0^{+\infty} (v^T e^{tA})(-C)(v^T e^{tA})^T\,dt.$$

Since $-C > 0$ and e^{tA} is nonsingular for all t, the integrand above is positive. Hence $v^T X v > 0$ and thus X is positive definite.

Conversely, suppose $X = X^T > 0$ and let $\lambda \in \Lambda(A)$ with corresponding left eigenvector y. Then

$$\begin{aligned} 0 > y^H C y &= y^H A X y + y^H X A^T y \\ &= (\lambda + \bar{\lambda}) y^H X y. \end{aligned}$$

Since $y^H X y > 0$, we must have $\lambda + \bar{\lambda} = 2 \operatorname{Re} \lambda < 0$. Since λ was arbitrary, A must be asymptotically stable. □

Remark 13.25. The Lyapunov equation $AX + XA^T = C$ can also be written using the vec notation in the equivalent form

$$[(I \otimes A) + (A \otimes I)]\mathrm{vec}(X) = \mathrm{vec}(C).$$

A subtle point arises when dealing with the "dual" Lyapunov equation $A^T X + XA = C$. The equivalent "vec form" of this equation is

$$[(I \otimes A^T) + (A^T \otimes I)]\mathrm{vec}(X) = \mathrm{vec}(C).$$

However, the complex-valued equation $A^H X + XA = C$ is equivalent to

$$[(I \otimes A^H) + (A^T \otimes I)]\mathrm{vec}(X) = \mathrm{vec}(C).$$

The vec operator has many useful properties, most of which derive from one key result.

Theorem 13.26. *For any three matrices A, B, and C for which the matrix product ABC is defined,*

$$\mathrm{vec}(ABC) = (C^T \otimes A)\mathrm{vec}(B).$$

Proof: The proof follows in a fairly straightforward fashion either directly from the definitions or from the fact that $\mathrm{vec}(xy^T) = y \otimes x$. □

An immediate application is to the derivation of existence and uniqueness conditions for the solution of the simple Sylvester-like equation introduced in Theorem 6.11.

Theorem 13.27. *Let $A \in \mathbb{R}^{m \times n}$, $B \in \mathbb{R}^{p \times q}$, and $C \in \mathbb{R}^{m \times q}$. Then the equation*

$$AXB = C \tag{13.14}$$

has a solution $X \in \mathbb{R}^{n \times p}$ if and only if $AA^+CB^+B = C$, in which case the general solution is of the form

$$X = A^+CB^+ + Y - A^+AYBB^+, \tag{13.15}$$

where $Y \in \mathbb{R}^{n \times p}$ is arbitrary. The solution of (13.14) is unique if $BB^+ \otimes A^+A = I$.

Proof: Write (13.14) as

$$(B^T \otimes A)\mathrm{vec}(X) = \mathrm{vec}(C) \tag{13.16}$$

by Theorem 13.26. This "vector equation" has a solution if and only if

$$(B^T \otimes A)(B^T \otimes A)^+ \mathrm{vec}(C) = \mathrm{vec}(C).$$

It is a straightforward exercise to show that $(M \otimes N)^+ = M^+ \otimes N^+$. Thus, (13.16) has a solution if and only if

$$\begin{aligned} \mathrm{vec}(C) &= (B^T \otimes A)((B^+)^T \otimes A^+)\mathrm{vec}(C) \\ &= [(B^+B)^T \otimes AA^+]\mathrm{vec}(C) \\ &= \mathrm{vec}(AA^+CB^+B) \end{aligned}$$

and hence if and only if $AA^+CB^+B = C$.

The general solution of (13.16) is then given by

$$\mathrm{vec}(X) = (B^T \otimes A)^+ \mathrm{vec}(C) + [I - (B^T \otimes A)^+(B^T \otimes A)]\mathrm{vec}(Y),$$

where Y is arbitrary. This equation can then be rewritten in the form

$$\mathrm{vec}(X) = ((B^+)^T \otimes A^+)\mathrm{vec}(C) + [I - (BB^+)^T \otimes A^+A]\mathrm{vec}(Y)$$

or, using Theorem 13.26,

$$X = A^+CB^+ + Y - A^+AYBB^+.$$

The solution is clearly unique if $BB^+ \otimes A^+A = I$. □

EXERCISES

1. For any two matrices A and B for which the indicated matrix product is defined, show that $(\mathrm{vec}(A))^T(\mathrm{vec}(B)) = \mathrm{Tr}(A^TB)$. In particular, if $B \in \mathbb{R}^{n \times n}$, then $\mathrm{Tr}(B) = \mathrm{vec}(I_n)^T \mathrm{vec}(B)$.

2. Prove that for all matrices A and B, $(A \otimes B)^+ = A^+ \otimes B^+$.

3. Show that the equation $AXB = C$ has a solution for all C if A has full row rank and B has full column rank. Also, show that a solution, if it exists, is unique if A has full column rank and B has full row rank. What is the solution in this case?

4. Show that the general linear equation

$$\sum_{i=1}^{k} A_i X B_i = C$$

can be written in the form

$$[B_1^T \otimes A_1 + \cdots + B_k^T \otimes A_k]\mathrm{vec}(X) = \mathrm{vec}(C).$$

5. Let $x \in \mathbb{R}^m$ and $y \in \mathbb{R}^n$. Show that $x^T \otimes y = yx^T$.

6. Let $A \in \mathbb{R}^{n\times n}$ and $B \in \mathbb{R}^{m\times m}$.

   (a) Show that $\|A \otimes B\|_2 = \|A\|_2\|B\|_2$.

   (b) What is $\|A \otimes B\|_F$ in terms of the Frobenius norms of A and B? Justify your answer carefully.

   (c) What is the spectral radius of $A \otimes B$ in terms of the spectral radii of A and B? Justify your answer carefully.

7. Let $A, B \in \mathbb{R}^{n\times n}$.

   (a) Show that $(I \otimes A)^k = I \otimes A^k$ and $(B \otimes I)^k = B^k \otimes I$ for all integers k.

   (b) Show that $e^{I\otimes A} = I \otimes e^A$ and $e^{B\otimes I} = e^B \otimes I$.

   (c) Show that the matrices $I \otimes A$ and $B \otimes I$ commute.

   (d) Show that
$$e^{A\oplus B} = e^{(I\otimes A)+(B\otimes I)} = e^B \otimes e^A .$$

   (Note: This result would look a little "nicer" had we defined our Kronecker sum the other way around. However, Definition 13.14 is conventional in the literature.)

8. Consider the Lyapunov matrix equation (13.11) with
$$A = \begin{bmatrix} 1 & 0 \\ 0 & -1 \end{bmatrix}$$
and C the symmetric matrix
$$\begin{bmatrix} 2 & 0 \\ 0 & -2 \end{bmatrix}.$$
Clearly
$$X_s = \begin{bmatrix} 1 & 0 \\ 0 & 1 \end{bmatrix}$$
is a symmetric solution of the equation. Verify that
$$X_{ns} = \begin{bmatrix} 1 & 1 \\ -1 & 1 \end{bmatrix}$$
is also a solution and is nonsymmetric. Explain in light of Theorem 13.21.

9. **Block Triangularization:** Let
$$S = \begin{bmatrix} A & B \\ C & D \end{bmatrix},$$
where $A \in \mathbb{R}^{n\times n}$ and $D \in \mathbb{R}^{m\times m}$. It is desired to find a similarity transformation of the form
$$T = \begin{bmatrix} I & 0 \\ X & I \end{bmatrix}$$
such that $T^{-1}ST$ is block upper triangular.

(a) Show that S is similar to

$$\begin{bmatrix} A+BX & B \\ 0 & D-XB \end{bmatrix}$$

if X satisfies the so-called **matrix Riccati equation**

$$C - XA + DX - XBX = 0.$$

(b) Formulate a similar result for block lower triangularization of S.

10. **Block Diagonalization:** Let

$$S = \begin{bmatrix} A & B \\ 0 & D \end{bmatrix},$$

where $A \in \mathbb{R}^{n \times n}$ and $D \in \mathbb{R}^{m \times m}$. It is desired to find a similarity transformation of the form

$$T = \begin{bmatrix} I & Y \\ 0 & I \end{bmatrix}$$

such that $T^{-1}ST$ is block diagonal.

(a) Show that S is similar to

$$\begin{bmatrix} A & 0 \\ 0 & D \end{bmatrix}$$

if Y satisfies the Sylvester equation

$$AY - YD = -B.$$

(b) Formulate a similar result for block diagonalization of

$$S = \begin{bmatrix} A & 0 \\ C & D \end{bmatrix}.$$

Bibliography

[1] Albert, A., *Regression and the Moore-Penrose Pseudoinverse*, Academic Press, New York, NY, 1972.

[2] Bartels, R.H., and G.W. Stewart, "Algorithm 432. Solution of the Matrix Equation $AX + XB = C$," *Comm. ACM*, 15(1972), 820–826.

[3] Bellman, R., *Introduction to Matrix Analysis*, Second Edition, McGraw-Hill, New York, NY, 1970.

[4] Björck, Å., *Numerical Methods for Least Squares Problems*, SIAM, Philadelphia, PA, 1996.

[5] Cline, R.E., "Note on the Generalized Inverse of the Product of Matrices," *SIAM Rev.*, 6(1964), 57–58.

[6] Golub, G.H., S. Nash, and C. Van Loan, "A Hessenberg-Schur Method for the Problem $AX + XB = C$," *IEEE Trans. Autom. Control*, AC-24(1979), 909–913.

[7] Golub, G.H., and C.F. Van Loan, *Matrix Computations*, Third Edition, Johns Hopkins Univ. Press, Baltimore, MD, 1996.

[8] Golub, G.H., and J.H. Wilkinson, "Ill-Conditioned Eigensystems and the Computation of the Jordan Canonical Form," *SIAM Rev.*, 18(1976), 578–619.

[9] Greville, T.N.E., "Note on the Generalized Inverse of a Matrix Product," *SIAM Rev.*, 8(1966), 518–521 [Erratum, *SIAM Rev.*, 9(1967), 249].

[10] Halmos, P.R., *Finite-Dimensional Vector Spaces*, Second Edition, Van Nostrand, Princeton, NJ, 1958.

[11] Higham, N.J., *Accuracy and Stability of Numerical Algorithms*, Second Edition, SIAM, Philadelphia, PA, 2002.

[12] Horn, R.A., and C.R. Johnson, *Matrix Analysis*, Cambridge Univ. Press, Cambridge, UK, 1985.

[13] Horn, R.A., and C.R. Johnson, *Topics in Matrix Analysis*, Cambridge Univ. Press, Cambridge, UK, 1991.

[14] Kenney, C., and A.J. Laub, "Controllability and Stability Radii for Companion Form Systems," *Math. of Control, Signals, and Systems*, 1(1988), 361–390.

[15] Kenney, C.S., and A.J. Laub, "The Matrix Sign Function," *IEEE Trans. Autom. Control*, 40(1995), 1330–1348.

[16] Lancaster, P., and M. Tismenetsky, *The Theory of Matrices*, Second Edition with Applications, Academic Press, Orlando, FL, 1985.

[17] Laub, A.J., "A Schur Method for Solving Algebraic Riccati Equations," *IEEE Trans.. Autom. Control*, AC-24(1979), 913–921.

[18] Meyer, C.D., *Matrix Analysis and Applied Linear Algebra*, SIAM, Philadelphia, PA, 2000.

[19] Moler, C.B., and C.F. Van Loan, "Nineteen Dubious Ways to Compute the Exponential of a Matrix," *SIAM Rev.*, 20(1978), 801–836.

[20] Noble, B., and J.W. Daniel, *Applied Linear Algebra*, Third Edition, Prentice-Hall, Englewood Cliffs, NJ, 1988.

[21] Ortega, J., *Matrix Theory. A Second Course*, Plenum, New York, NY, 1987.

[22] Penrose, R., "A Generalized Inverse for Matrices," *Proc. Cambridge Philos. Soc.*, 51(1955), 406–413.

[23] Stewart, G.W., *Introduction to Matrix Computations*, Academic Press, New York, NY, 1973.

[24] Strang, G., *Linear Algebra and Its Applications*, Third Edition, Harcourt Brace Jovanovich, San Diego, CA, 1988.

[25] Watkins, D.S., *Fundamentals of Matrix Computations*, Second Edition, Wiley-Interscience, New York, 2002.

[26] Wonham, W.M., *Linear Multivariable Control. A Geometric Approach*, Third Edition, Springer-Verlag, New York, NY, 1985.

Index